From Evolution to God:

An Integration of Faith and Science

ROBERT DEPAOLO

"From Evolution to God: An Integration of Faith and Science," by Robert DePaolo. ISBN 978-1-63868-065-9 (softcover).

Published 2022 by Virtualbookworm.com Publishing Inc., P.O. Box 9949, College Station, TX 77842, US. ©2022, Robert DePaolo. All rights reserved. No part of this publication may be reproduced, stored in a retrieval system, or transmitted in any form or by any means, electronic, mechanical, recording or otherwise, without the prior written permission of Robert DePaolo.

Books by Robert DePaolo

Society Reconsidered: A Debate on the Issues of Modern Times

Hominids: A Perspective on Human Biosocial Evolution from the Treetops to the Renaissance

American Evolution: The Flaws and Promise of a Unique Society

Evolution, Information and Personality: Toward a Unified Theory of the Psyche

Resurrection

Bundy: A Clinical Discussion of the Perfect Storm

This book is dedicated to four parties: A sister and wife battling illness with courage and humor, and the scientists and theologians who have sought to interpret for us all the wonders of God and nature.

Table of Contents

Introduction

THIS BOOK DISCUSSES TWO GUIDING PRINCIPLES of human experience; faith and science, based on the assumption that there is a fundamental connection between the two. In recent times science and faith have been separated by sentiment, language, and law. However, the relationship between the two disciplines seems to have been more intertwined than mutually exclusive over the course of human history. That is because, arguably, both have been virtually identical in their purpose.

Both science and religion have served, and continue to serve, human needs. Both have reflected and been guided by human cognitive and emotional traits resulting from the evolutionary development of the human brain. Both continue to provide emotional relief by conveying information to the mind and heart, in the process offering anxiety-alleviating control over an environment replete with duress, particularly in terms of the human capacity to conjure up sadness based on internal appraisals of experience.

Ultimately, both derive from the same source - nature. Religious guidance comes through belief in all-powerful beings who embody all aspects of the natural world, including its forces, its provisions, its destructive power, and its capacity to nurture. Scientific guidance comes through ideas provided by human observation.

There has always been a question as to whether there is a significant distinction to be made regarding the relative validity of the scientific method vs. the belief that God has control over the natural world. Beyond that is the question of whether there exists a neuro-cognitive component resulting from evolution that produced the scientific method and at the same time led virtually all human societies to believe in God - or at least in some causal entity outside themselves.

Despite the mystical, spiritual underpinnings of most religions, the belief in God as a built-in element of human experience has existed throughout the course of history. Prophets in the Old and New Testaments believed God was housed within the human heart. For example, Jesus emphasized that good and bad was within us, that free will (as opposed to absolute divine determinism) was entrenched in human nature, that God does not orchestrate all events, triumphs, and tragedies.

Notwithstanding the devotion to various gods over time most, if not all faiths have conveyed the idea that morality is subject to human control, that God could only judge fairly if good and bad acts were a function of human intent. In that sense, there has always been an interactive relationship between God and man that is less subservient than contractual. God rewards us for moral acts and punishes us for immoral acts.

Yet His decisions have not been arbitrary. While no religion that I'm aware of espouses equity between God and man (there would be no reason to worship an entity no more powerful than us), all seem to require choice for the concepts of sin and morality to have meaning. That means the internal experience of Homo sapiens is a significant factor in how we conceptualize the nature of God.

In the final analysis, cognitive and emotional factors giving rise to free will must be attributed to brain functions. The human brain is concrete, yet symbolic, oriented in the present, yet futuristic in its breadth. It is the imaginative organ from which thoughts, predictions, emotions, symbols, beliefs, the fear of punishment and the anticipation of reward derive. That means the way the human brain evolved has deep relevance for the development of religious beliefs.

If the Bible had been written in an anthropological context one chapter (perhaps in Genesis) might state that God could only have presented his laws to Homo sapiens because, despite His creation of all beings, only man had the language capacities and emotional depth to understand and act in accord with those laws. Furthermore, since those capacities appear to have developed over the course of hominid evolution, it makes sense to assume religious beliefs arose gradually, and in accord with the cognitive capacities of archaic and ultimately, modern humans. There are suggestions of that in The Old Testament, albeit conveyed indirectly.

Obviously, Genesis did not delve directly into human evolution, but with a little imagination much of the text can be viewed as derivative of both Darwin and Abraham. For example:

GENESIS 20. And God said: Let the waters bring forth abundantly in the waters the moving creatures of the bath life and fowl that may fly above the earth to the firmament of heaven.

GENESIS 21. And God created great whales and every living creature that moveth, which the waters brought forth abundantly after their kind, and every winged fowl after his kind. And God saw that it was good.

GENESIS 22. And God blessed them, saying, be fruitful and multiply and fill the waters in the seas and let fowl multiply to the earth.

GENESIS 24: And God said; Let the earth bring forth the living creature after his kind, cattle and creeping things and beasts of the earth after his kind, and it was so.

GENESIS 25: And God made the beast of the earth after his kind and cattle after their kind and everything that creepith upon the earth after his kind and God said, all was good.

In these passages, there is a succession that parallels the evolution of life forms: first the sea creatures, then land creatures, then man. Each passage alludes to the emergence of all forms of life - as if describing the genetic distinctions between various species. Interestingly, the author (s) of Genesis pointed out that God did not create man first.

Moreover, there is no indication that the earlier creation of other animals was to support man. Sea creatures and creeping animals would have had no central role in man's existence. It's just that in the phylogenetic sequence God made them first.

Of course, all of that could have been written by writers who simply knew something about various life forms during the time Genesis was composed. Still, a later chapter in Genesis also seems to wax evolutionary. According to the text in Genesis 1:27,) while man was created in God's image, he is initially portrayed in Eden as having no shame, despite being naked.

What could that have meant? If man was created in God's image, he would presumably have had superior cognitive abilities. That would ostensibly lead to self-awareness and a capacity to feel shame. Unless by "image" the Old Testament is referring solely to physical appearance, the naivete of Adam and Eve during their period of innocence is hard to explain.

At the time Genesis was written, being naked was an extremely humiliating experience. It still is today for all adherents to the Abrahamic religions. That makes the scene in Eden most interesting. Why the absence of shame? Various writers have addressed that question, perhaps most notably biblical scholar John Golding in his book, *Genesis*. The explanation was that prior to committing sin - by eating from the tree of knowledge, it wasn't possible to feel shame. This suggests guilt could only have arisen from a sinful act.

On the other hand, if there was a capacity to look inward and process the good or bad of one's own behavior, why

4

wouldn't shame have been housed in the mind of Adam and Eve from the moment of their creation?

Indeed, it seems two questions are involved in this episode. First, did they feel shame after partaking of the forbidden fruit? Second, were the first two humans wired in a way that made shame possible?

There are ways to address these questions in both a religious and anthropological context. One possibility is that Adam and Eve were without shame because they were the only two people alive and could not have been subject to shame by others. In other words, sin, which is some function of social mores, had not been "invented." Therefore, Adam and Eve had no broad social frame of reference by which to judge themselves and could not understand, let alone fear the possibility of ostracism.

Another possibility is that an incapacity to feel shame reflected a moral naivete, that is, an incapacity to feel shame due to a lack of cognitive depth that would later facilitate self-awareness and set the stage for feelings of guilt and shame

In that context, God would appear to have regarded Adam and Eve as higher beasts who had not quite reached an introspective threshold - beasts he hoped would go forth and multiply. Being naked would have provided a provocative stimulus in that respect.

If shamelessness could be attributed to the fact that the first two humans lacked the capacity to feel shame, it could be viewed as a parallel to the evolution and development of human cognition which proceeded from the primitive mammalian brain (with its id-dominated, survival - based proclivities) to a neo - mammalian and ultimately human brain, which expanded in the frontal cortex and provided a capacity for self-reflection and, eventually, for guilt.

That assumption is, to an extent, supported by a study by Koenigs and Motrin at the University of Wisconsin - Madison on psychopaths. The subjects, who were all diagnosed as being sociopaths, exhibited diminished connectivity from the frontal lobes to other brain sites.

It is also significant that in Genesis the capacity for self-consciousness arose after the couple was persuaded by Satan to partake of the tree of knowledge. That suggests pride, a self-reflective process, and the increasing quest for knowledge, which is a function of the cortically driven curiosity drive, had not yet existed when Adam and Eve first appeared on the scene. That is reminiscent of the transformation from an external to an internal cognitive capacity that eventually enabled humans to process feedback resulting from their own actions.

If presented in both biological and religious terms, it could be interpreted as a warning from God. To wit: *Don't let your brain growth and concomitant increase in intelligence go to your head*, (pardon the pun). God's admonition in Eden seemed to imply that the quest for knowledge was not all for the good. Indeed, the message conveyed was that unpredictable outcomes could result from substituting human epistemological greed for submission to the Lord, and that self-satisfaction and pride resulting from resolution could turn out to be destructive (a sentiment that has been reflected in modern day protests against cloning, nuclear proliferation and pollution due to industrialization).

All the above presumptions are of course speculative, yet there do seem to be parallels, rather than absolute conflict between the basic tenets of Judeo-Christianity (as well as other faiths) and the tenets of modern science.

In a sense, the science/religion dichotomy is irrelevant. The history of both bodies of thought has chronicled Homo Sapiens' psychological journey through time, during which our species became increasingly adept in the areas of critical thinking, moralizing, storytelling, interacting and concluding. Since, as God intended, man has a central role in outcomes, it stands to reason that science and religion might be congruent rather than mutually exclusive bodies of thought.

Whether one accepts that Genesis has an evolutionary tinge, the fact is, religion and science present a functional mirror image when examined closely, particularly in that both serve the purpose of survival.

As a scientifically derived process, evolution (natural selection) operates by favoring aspects of behavior and physical structure that provide advantages to the individual and the group. The Ten Commandments and the tenets of the Islamic, Buddhist and Hindu faiths are also geared toward survival of the individual and the group, The rules in all religions prohibit killing, stealing, infidelity, bearing false witness and other actions that foment social conflict and threaten to undo the social solidarity needed to secure resources and provide for defense against rivals and predators and as galvanizing actions when faced with natural disasters.

In the first nomadic tribes, whose lifestyle characterized most of our time on earth, having an adequate number of members to hunt, fish, make tools, propagate, provide protection, scout, explore and care for children was paramount. This was especially true because our nomadic ancestors had transient cultures. They had to remake tools in each location and find new resources in each new locale. It was a highly redundant lifestyle.

Beyond that, as Yuval Harare discussed in *Sapiens; A Brief History of Humankind,* due to the rigors of their existence, old, fragile, and ill members were often left behind, at least until caves were discovered. They most certainly cared about fragile members - even Neanderthals buried their dead with compassion and in apparent belief in the afterlife. But nomadism precluded stability that could lead to progress and invention.

If the concept of sin existed in early nomadic human groups, it might have been the result of the enhanced depth of emotion arising from brain expansion, whereby the associative cortex enabled them to embellish feelings through an increasing associative range of words and memories. That alone might have led to a merger between the survival instinct and belief in a deity.

There are factors beyond the need for group solidarity and family integrity that are common to both biology and faith and improve survival capacities. The concepts of loyalty, fidelity, love, compassion, the quest for prosperity (food and drink being necessary to group survival), social cooperation and competition are reflected in biological/evolutionary as well as religious principles. Whether based on spiritual or biological motives, both faith and science appear pragmatic mechanisms emanating from the mind. In other words. both faith and science work.

Does that mean religion is an adaptive cognitive trait, just as necessary to our species as the capacity for upright walking, language, and toolmaking? In earlier times, and even now, one could argue that it is, despite the drift toward science as a prime reference point in modern times.

There are differences as well, the main one being that science relies on mathematical proofs. Faith has no standard deviation. It can't be measured through correlation coefficients or T scores. For that reason, hard core atheists have argued that religion is no longer necessary. One of the more prominent critics of religious worship has been Christopher Hitchens. In his book *God is Not Great; How Religion Spoils Everything*, he argued that religion is not only pointless but destructive.

But that is not true, either in a historical or moral context. Aside from the argument here, that given the structure and functions of the human brain, some form of worship and/or belief in transcendence is unavoidable, there are many instances in which humanistic, philosophical, epistemological, altruistic, and even political principles derive from religious tenets.

For an atheist to make their case it seems they would have to engage in guesswork on how human nature would have evolved in a complete absence of religious thought. Since there has never been a time when some sort of divine worship did not exist in human society, Hitchens' argument seems questionable. The fact is Homo sapiens needed religion to survive. That might be why our species has been so reluctant to abandon faith.

History seems to bear that out. There was no Arabic numeral (decimal) system prior to the 9th century. Even for a time after that, math discoveries were attributed to God's intervention. For all their innovative work in the areas of geometry, science and anatomy, Greek philosophers Pythagoras, Aristotle, Archimedes, and Galen always assumed knowledge of natural phenomena derived from a deity. As late as the 17th century, Isaac Newton, whose invention of calculus contributed greatly to modern science, believed the hand of God was involved in mathematical structures and all aspects of nature.

However, it wasn't as if early religious and philosophical thinkers operated in a subjective vacuum. They had (thanks to Aristotle and others) a pre-mathematical method that was reasonably objective. It was a semantic approximation of math known as the syllogism.

A syllogism is an irrefutable statement of fact that does not require mathematical proof. The most typical example is seen in the axiom, if A is greater than B and B is greater than C, then A must be greater than C. There are no numbers in that statement, yet it is unquestionably true - even more so than many scientific conclusions, which are based on probabilities rather than absolute certainties. Many philosophers and clerics used that method prior to the 9th century, and while the method sometimes resulted in questionable conclusions, particularly regarding ideas put forth by Aristotle, it was often accurate. Its precision depended on the validity of the initial premise. The syllogism did not surpass mathematical analysis, but it is not as if thinkers like Aristotle, Plato, Jesus of Nazareth (whose parables in Matthew 6:28 and Matthew 13:31-2 were replete with syllogistic reasoning) St. Augustine and Thomas Aquinas were operating in the dark.

It is also important to point out that science - which begins with a hypothesis rather than a statistic, has also yielded incorrect conclusions. Consumption of wine has been deemed both healthy and dangerous at various times. The same is true with caffeine, carbohydrates, and proteins. Over time the "normal" range for blood sugar levels has been raised. Standards for obesity have been lowered and studies "proving" children in Head Start programs had an increase in IQ have been disconfirmed.

Thus, it seems reasonable to conclude that religion and science derive from similar cognitive traits unique to our species, and because we can be fallible, so can our ideas and conclusions.

Rather than dismiss either fallible science or non-verifiable religion, it makes sense to assume they are adaptive cognitive patterns that over time have been of great benefit to our species. Indeed, because of their common human connection, both could be described as branches on a tree of knowledge that are tethered to a common trunk.

The task lies in determining the nature of that foundation. One place to start is with the notion that God and his moral message can ultimately be found within us.

CHAPTER 1:
ORIGINS

GOD COULD NOT HAVE IMPARTED ARTICLES of faith and law unless people were capable of understanding and abiding by those laws. No matter how obedient the flock, the tenets had to reach both the intellect and the emotions. Through evolution, the brain/mind complex of Homo sapiens developed a capacity to understand right and wrong and to have emotional reactions to moral and immoral acts.

That would have included an appreciation of not only what defines prosocial and antisocial behavior, but also criteria by which to evaluate the differences. Whether in a socio-biological or religious context, actions can only be judged according to their proportionate impact on the individual and the group. When in (Deuteronomy 19:21) the Bible refers to the principle of 'an eye for an eye' and a 'tooth for a tooth' it is not a statement of revenge but one of proportion. It meant *only* an eye for an eye and *only* a tooth for a tooth. This implies that any punishment that supersedes the impact of the transgression is itself sinful.

The tendency to engage in proportional thinking seems to have characterized our species from the outset. If not for that capacity, it would be hard for us to function within a structured society. We would have no morals and no laws.

This trait was entrenched in the earliest pantheistic and pagan religions, as well as the legal system developed by Hammurabi of Babylon and Darius of Persia, and it appears to emanate from the neural organization of the human brain. That is important with respect to the influence of human cognitive evolution on religious beliefs.

Beyond a capacity for proportional thinking, the brain developed a capacity for imagery and other internally generated experiences. The ability to look within for purposes of self-evaluation seems to have resulted from expansion of the frontal cerebral cortex, which the research of neurocytologist J.M. Fuster demonstrated, has vast connections to almost all other brain sites. Despite having ubiquitous influence on behavior, it has no specific function. Its neuro-functional breadth and pervasive influence allows it to operate as a super-observer and regulator of memories and actions. That enables humans to comprehend good and bad aspects of one's own actions as well as the behavior of others. Through the vista of frontal expansion, we can "see ourselves"- as though looking in a mirror.

In that context, it might be helpful to discuss how such neural functions developed during human evolution, and at what point in human development Homo sapiens was able to accept God into his experience Homo Sapiens' evolutionary path toward religiosity might have begun with the advent of language, which is, to a large extent, regulated by neural circuitry extending from the parietal to the frontal cortex. Evidence suggests the parietal cortex, specifically Broca's and Bernice's areas, are concerned with expressive and receptive language functions, while the frontal lobes provide more fractional, inhibitory, internalized language functions that provide the capacity for self-guidance, contemplation, and guilt. In somewhat colloquial terms, the frontal cortex could well be described as a feedback circuit processing all that we are and all that we do. There are various theories of how and when human language originated but given the structure of the brain, a certain evolutionary timeline seems plausible.

The first upright walker (hominid) arrived on the scene roughly four million years ago. This creature was given the name Australopithecus. His brain was not significantly larger than a chimpanzee's. However, due to his upright posture, an anatomical realignment occurred, featuring a vertical shift in the positions of the spinal column, skull case and pelvis, which took on a concave shape to provide a comfortable mechanical resting position for the spine. These permanent realignments became important to both human anatomy and human perception

Chimps and other great apes occasionally walk upright, and this can enhance their frontal range of vision. Being able to spot predators and survey distant landscapes, even if only briefly, can provide information conducive to survival. In that moment the chimp can summon a degree of contemplation because he is, by definition, looking ahead at something he "could" encounter at some future point but does not have to deal with presently. In other words, distance vision was a possible starting point for what we now call the imagination. It likely led to anticipatory thought and to a rudimentary ability to make predictions.

The chimp is a quadruped, so this faculty could never have developed to the extent it did with the first hominids, who were able to look ahead for extended periods of time. Once upright walking became permanent it became possible for bipeds to view and ponder both proximal and distal stimuli. Distal perception does not typically require an immediate response. Thus, an image could ruminate in memory, which would have facilitated a longer attention span and set the stage for what we now refer to as perspective.

Perspective was probably not available to the Australopithecines except perhaps in very primitive form, but while their brains were not significantly larger than a chimp's, a protrusion of neurons developed in a middle section of the cerebral cortex known as the parietal lobe. This area contains a motor strip that regulates movement of the fingers, mouth, tongue, and other small muscle groups. Expansion of that section would have fine-tuned responses in the hand, fingers, lips, and tongue. That would have led to refinement of vocalizations through vocal, as well as digital streamlining, and set the stage for a linguistic revolution among primates verbally, and through manual gestures.

As their posture became upright, muscles in the diaphragm had greater spatial volume. That enabled the hyoid bone in the throat to lower its position and serve as an echo chamber so that vowel sounds could be more prominent and of longer duration. While the sounds made by a chimp are choppy, the sounds of the first hominids were possibly more melodic and expansive.

The increase in sound duration, volume, and phonetic detail, would have enabled them to convey more information in vocalizing. That would have been augmented by manual gestures, such as pointing and waving to convey spatial understandings.

The first hominids were likely on the road to a prototype human language but were not quite there yet. Their communicative style, perhaps had a sing-song quality with rhythmic features, including extended, scaled, rather than monotone howling, and perhaps more elaborate accentuation. While still primitive, those vocal enhancements improved the connection between emotion and expression.

Music not only soothes the savage beast, it can also facilitate expression of a wider variety of moods. Thus, it is possible emotions such as sadness, elation, anger, and affection were also becoming more pronounced.

Meanwhile, bipedal movement created the luxury of enhanced long distance travel skills, not just because of enhanced vista and predictive capacities but because an upright posture provides less of a target for the sun's rays, thereby making the hominid less affected by heat. The hominid was able to travel long distances without becoming dehydrated. That, accompanied by changes in cognitive style and capacity for enhanced vista, enabled hominids to become more efficient migrators.

Traveling proficiency was not always advantageous. Being on the move meant having to wait longer, sustain motivation longer, perhaps even learn to postpone basic appetitive concerns in anticipation of future bounty - although that would have been mitigated by the fact that they would not have to eat or drink as frequently due to their anatomical insulation from the heat of the sun.

The first species of hominids, Australopithecus, Afarensis and Robustus, were not migratory. It appears they spent most of their time on the forest floor and, when faced with predators, or in the process of building nests for their young, they reverted to the trees.

They did not have the patience, long term cognition, or attentive skills of subsequent hominids such as Homo habilis, Homo erectus, Homo ergaster and Homo Heidelbergensis. In one of those random evolutionary occurrences, patience made possible by vista, extended phonics (which made auditory signal processing lengthier) and heat resistance resulted from a unique kind of brain expansion. There was a proliferation of neurons in the frontal cortex. That is significant in terms of the relationship between primate brain structure and function.

If Darwin's theory of natural selection is even partly valid (this writer considers it only partly accurate) then the distribution of nerve cells in the brain of a given species will tend to reflect how it adapted to its habitat.

Primates have always been spectacularly visual creatures. They are the most colorful animals on the planet which correlates with their superb capacity for visual perception and visual discrimination. They must have excellent depth perception with which to navigate in the trees without risking injury. As a result, the most prominent circuits in primate brains are in the occipital lobe, a posterior segment of cortex concerned with visual processing. A prominent occipital lobe remained a feature of the hominid brain even after primates left the trees for the open savanna.

That trait changed only a bit with early hominids, but really took a back seat with the arrival of archaic human populations. The brains of archaic humans featured massive expansion of the parietal and frontal lobes of the cerebral cortex. The frontal lobe housed extensive inhibitory circuits along with increased myelination (a fatty layer of brain tissue that enables the brain to tolerate prolonged stimulation, thus devote more time to problem solving). Through that process, hominids were able to whittle down movement with finer precision in the hands, tongue, and fingers. It appeared fine motor skills were being selected by nature. That created momentum for toolmaking and led to enhanced attention faculties. These frontoparietal circuits provided greater capacity to pause between perception of stimuli and response selection, which led to deeper analysis of variables and an improvement of problem-solving skills.

More important, is that the newest cluster of frontal neurons were less specifically functional. As has been noted in the famous case of railroad worker Phineas Gage, significant damage to the frontal lobes doesn't necessarily interfere with specific brain functions. It appears these circuits did not develop to support memory, sense perception or motor faculties. Instead, they feature a massive increase in neurons, many of which are inhibitory. The capacity to put inputs on hold, not just because of inhibition (which is a function of neuro-chemical properties) but also because the sheer mass of the frontal cortex mitigated against rapid, reflexive behavior.

Frontal lobe expansion provided other cognitive advantages. Acting as a voluminous filtering mechanism, this brain section allowed more data to be processed and compared, both from the environment and from stored memories. In effect, frontal expansion led to conceptualization of the natural world and to the advent of internal experience. That allowed late hominids and early humans to take in more information, develop elementary categorizing capacities, draw conclusions about the similarities and differences between and among elements of nature and the vocalizations of members within their own group, including sounds within their language structure.

The impulse-driven cognition of primates and early hominids was gradually replaced by deliberation. That provided a more holistic sense of the world, so that events, circumstances, and one's own actions could be categorized and evaluated. It created an extended level of consciousness and of the conscience as we understand it today.

The cognitive/ linguistic stage was set. Early humans were able to search for, accept and understand the nature of God. It is impossible to determine which hominid group first became religious. However, the ability of Homo erectus to make and use fire suggests he was able to override the fear of fire (which is typical of all animals) and view it in a creative context. Such a cognitive conversion would have required a re-configuration of experience, which meant a re-labeling of stimuli, which suggests a rudimentary capacity for categorical reasoning.

Still, there is no evidence regarding which group first had religious thoughts. Paleoanthropology deals mostly with bones, genes, and artifacts. While scientists can determine what species a skull belongs to with some accuracy, they cannot look inside to see what thoughts and perceptions might have arisen from within that skull. In any case, the ability of later hominids to take the long view of both the terrain and the self might have created a scaffold for the emergence of faith.

Did Homo erectus or his contemporary, Homo ergaster worship a God? Were ancient cave paintings in the islands around current day Australia indicative of a spiritual inclination? It is difficult to say, but it seems possible these groups had the mental faculties and perhaps the neuro-behavioral impetus for spirituality.

Later, with the origin of Neanderthal it appears religion was in play. What form this took is unknown, but one can extrapolate from archeological discoveries in the caves of southern Europe and the Middle East. Neanderthal buried the dead with supplies - not unlike the Egyptian custom of placing foods and other staples in the tombs of pharaohs. The Egyptians wanted to ensure the deceased was able to continue in an afterlife, which suggests they believed life on earth was merely one part of a dual existence. It seemed they either did not believe death was a finality, or perhaps they could not face the prospect of a revered figure being taken from them. One suspects it was a combination of both.

With respect to the Neanderthals, it is difficult to determine what emphasis was in play because they did not, perhaps could not provide historical accounts through any sort of symbol system. However, they did descend from one of the advanced, hominid lines, which meant they had vista, a degree of perspective and a semblance of self-awareness. Also, with a capacity to compare the behaviors, expressions, and status of individuals in the group there was undoubtedly adherence to a social hierarchy. Those highest in rank would have been considered more valuable, their loss more costly to the group. There would have been reluctance to "let them go," That, along with capacities for futuristic thinking and deliberation might have led to belief in an afterlife.

As author Barbara King suggested, being able to conceive of an afterlife would have led, along with vista and the ability to ponder, to a description of the place in which the deceased would end up residing. It might seem presumptuous to attribute such advanced religiosity to a species considered less advanced than our own. On the other hand, cognition entails an associative sequence in which one thought usually leads to another. Just as neurons in the brain branch out in spreading influence, so do the experiences represented in those connective fibers. If there was some language capacity, impressions, and observations from one member of the Neanderthal group would have been communicated to others. That could have created a social chain reaction, leading to proto-religious customs and a central set of rituals and beliefs that galvanized the group.

What does this mean, as per the relationship between faith and evolution? In her book *A History of God,* writer Karen Armstrong wrote that since all human groups have at one time or another worshiped Gods, it might be an inborn trait. In that context, she has referred to our species as Homo religiosis.

Given the neurological makeup, travel habits, emerging linguistic skills, and advanced capacity to feel emotion in more intense and varied ways, human beings and at least some of our predecessors seemed destined to worship God. Thus, it seems the idea of God and creation of religious systems might not have arisen artificially out of ignorance about how nature works but might have been built into us through evolution in terms of what Darwin called a conversion.

If the notion of God seems to have been planted in our minds as well as our hearts from the outset, we had no choice but to move forward with religion as a guiding principle. Does this give credence to the notion that God is within us?" If so, it would appear the debate on whether to choose between science or faith is moot.

Obviously, there is a difference between what clerics mean by the phrase "God is within us" and the biological implications referenced here. While a strictly religious interpretation has the Lord placing an engram of sorts into the soul, creating the capacities for goodness, virtue and faith, the idea here is that nature created the potential for a religious template in the neurological structure of the human brain. In any case, the distinction between metaphor and neurology, the spiritual and the natural, might be irrelevant.

Nature gave us a brain capable of anticipating death; one that is designed to control life's circumstances, and a brain that is highly emotional. Therefore, if one cannot accept death, it becomes necessary to conjure up a solution to that problem. The solution will be active, which means there will be some attempt to deny finality and act in ways to override death. The solution will coincide with the laws of associative thought, with one link leading to the next. The logical solution would be to assume the deceased is still alive in some sense and exists in some other place. If people decide he does exist, he will be presumed to function as he did while living, which means he would need companionship, food, and other supplies.

Once that precept is established, the parameters of the 'place' will be further defined. If there is a specific place, the assumption might be that others are in that place. If a posthumous society does exist, it should reflect the same social structure as on earth, which means there will be leaders and followers.

The leader will have to be an entity above all others. This would be reflective of hierarchical social instincts typical of both humans and primates. Within that rubric there would have to be alpha entities, including Gods and spirits.

The associative thought process would inevitably lead to religious ideation. It suggests, as referenced above, that God could be so entrenched in the minds of man as to be derived from evolutionary neuronal/cognitive propensities.

This is not meant to imply that the idea of God is some sort of neuro-behavioral quirk. It simply means there is no way to separate God (or transcendence of some sort) from the makeup of man. Moreover, because Homo sapiens is a product of nature, there is no way to separate God from nature. In effect, that means religion must entail worship of a oneness - a Supreme Being somehow distinct from, yet woven into the natural world.

CHAPTER 2:
THE SOCIAL ANIMAL

AUTHOR ELLIOT ARONSON WROTE that human beings are quintessentially social animals, but the core reasons why are somewhat complicated. One prominent theory, proposed by R.M. Dunbar, holds that animals with large brains (relative to body size) tend to be more social; for example, birds and primates. It makes sense, not so much because such creatures need social support and cooperation to survive, but because having a large brain enables an organism to make finer perceptions and recognize sensory patterns and differences in all objects, including members of their group. In other words, greater social interest arises from greater social recognition which in turn is made possible by greater brain volume.

Once a social template for enhanced social perception is established through increased brain volume, the organism will begin to interpret experience in social terms. In the case of primates and humans, that will lead to social dependency. It will involve social comparisons, social loyalties, tribal affiliations, and rivalries. Most cognitive schemes will be concerned with what others are doing, feeling, achieving, and plotting. Such an embellished social focus will often transfer to elements of the outside world, even if those elements are inanimate. Storms will be given feminine names. Ships will be named for generals and presidents, pets will be named for cartoon characters and planets will be named for gods, goddesses, and other mythical figures.

While that level of social perception was relatively new in evolutionary terms, one aspect of social perception has always existed, even among creatures with small brains. For example, the tendency among most animals is to organize their group into hierarchies, typified by dominant and submissive males and females and rankings in between. Rank will determine access to resources, mating options, as well as decision-making powers. Male dominance is a particularly influential factor.

A study by I.S. Bernstein demonstrated that among apes dominant males often exhibit two opposing characteristics; a tendency toward brutality and a tendency to produce a calming effect through their protective actions. Obedient members are often treated favorably while rebellious, contentious types are typically punished or ostracized – unless they succeed in deposing the alpha male.

It would appear no matter how large the brain, there will be adherence to those basic socio-biological dynamics. Rank can be refined, altered, and finessed, but in the final analysis, it can never be completely abandoned.

This is especially true of our own species. The question is whether such an ingrained social mindset was inserted into religion through evolution. In other words, was religion derived in part from a primal need for a social hierarchy?

Another aspect of the social interest phenomenon is seen in one person's perception of another, particularly as pertains to genetic closeness. Most species treat kin differently than they treat strangers. Our closest genetic cousins, the chimps, can be exceedingly brutal toward strangers. Even more telling is that chimp males will sometimes kill infants sired by another male after taking interest in a female. As discussed earlier, social behavior is often related to genetic closeness.

What does that say about faith and human evolution? Perhaps it can help explain the tribal aspect of some religions, as seen in use of the term 'chosen people' and the Islamic tenet; my cousin and I against my enemy, my brother and I against my cousin.

For most creatures, social rank is unconditionally accepted. The silverback male rules the gorilla troop. The ram that wins the lengthy battle against its opponent when vying for access to a female will always end up mating. The female always chooses the winner, while the loser, if still alive, will walk off quietly. The same is true of elephants, lions, and apes.

Yet, due to enhanced social perception, rank order in human societies is more complicated. Humans can expand and modify the concept of rank to boost our own status. We might lose a fight, earn less money, attain less fame and be less popular than others, but still be able to come up with a rationale (often bolstered by defense mechanisms) that keeps us high up on the totem pole. In fact, psychoanalyst Alfred Adler based his entire approach to psychiatry on this tendency, which he described as a lifelong conflict between feelings of inferiority and superiority.

One reason for this tendency lies in our language capacities. In his book *How the Mind Works*, neuroscientist and writer Steven Pinker expressed a belief that the human brain has circuits devoted to specific functions - much like a computer program, and that language is one of those programs. His interesting theory has met with criticism by some, who view language as emanating from other functions, such as fine motor control, hand gestures and imitative sound making. Regardless of which explanation is most accurate, language clearly is a powerful influence in human experience. For example, Psychologist Albert Ellis has written that our emotions correlate with our language expressions and appraisals. This is also true regarding what others say. For example, a 2019 study in Psicologia: Reflexao e Critica showed that depression in children correlates with habitual use of depressive language by mothers.

Such influence can extend to perception. Humans are the only creatures on earth who can defend themselves against reality. Through language, we can distort, deny, rationalize, suppress, and embellish experience. Oftentimes, we respond to an internal linguistic appraisal of events rather than the event itself. Indeed, that is a foundational component of the psychological theories of Richard Lazarus, Alexander Luria and Ivan Pavlov.

For example, a significant amount of research by Richard Lazarus has shown that language appraisals regulate mood and behavior. One result of that capacity is that one's sense of self and position on the social hierarchy can be altered virtually at will. A weakling can transform himself into a god. A man lacking in charm can presume he is extraordinarily charming and will tend to blame females who cannot acknowledge that. With enough "self-programming", he can even take this to a sociopathic level by seeking dominance against women in order to bolster his contrived grandiosity.

When it comes to the rules of social organization, some sort of authority must be deemed responsible for governance, and for exacting consequences. Therefore, if, as exemplified by the capacity for self-embellishment, the hierarchy is potentially fluid, the question involves reaching a consensus on who has the highest status.

It is an important consideration, because artifice, ambition, rivalries, and politics can render the hierarchical process so vague as to foment severe social conflict. Human beings might, through various ideologies, opt for equality as a social and political staple and come to believe no one is better than anyone else. But that is a political, rather than an anthropological construct.

History has shown that most people find it unsettling when one person or entity seeks absolute control over others. The idea that some group or individual has elite status is a seedbed of resentment from which many revolutions have arisen.

One solution to that problem originated in western democratic systems with the establishment of God as the ultimate authority. The idea was that if God, rather than a tyrant is solely responsible for bestowing rights to all people, those rights would have to be considered inalienable. God can't be deposed.

In that context, it is not surprising that a belief in God was employed for both religious and political reasons over time, as a means of maintaining social stability and preventing mass conflict. Because the idea of God as ultimate authority was grounded in both politics and religion it came to be considered both lawful and virtuous.

On the other hand, history has also shown that God and the state can be finessed into a fusion between God and king and become oppressive - as occurred with the notion of Divine Right, the Spanish Inquisition, and the witch trials in Salem.

Due to the evolution and refinement of human language, religious worship has been reshaped in various ways. Indeed, a central question regarding faith has always been: how much of religion is concocted by man and how much is imparted by God?

An even more interesting question comes out of the Deist school, which many of America's founders referenced in developing the Constitution. Thomas Jefferson, George Washington, Benjamin Franklin, and many others were Deists. As such, they believed God and nature were intertwined, that God did not control all circumstances via some sort of mysterious, unknowable plan. They reasoned, if that were true, and if all actions and outcomes were orchestrated by God, humans could not fairly be judged. Immoral acts would have to be viewed as a function of God's will, rather than man's free will. In that scenario man would be merely a puppet with no soul, who simply carried out predetermined decisions.

Deists also believed that in creating the universe God would have complete knowledge of how it worked and would unfold, therefore had no choice but to let it run its course. At first blush, that seems to suggest there is common ground among Darwin's theory of natural selection, the uncertainty-driven world of Werner Heisenberg, Richard Feynman's Sum of Histories theory, the heliocentric model of Copernicus and the nature of God; that is, a gigantic set of connections providing a vastly inclusive 'theory of everything.'

In that context, it isn't surprising that many post-Enlightenment philosophers, including Deists, believed God and nature were synonymous. It was an idea flush with explanatory potential as well as powerful ancient roots, and one worth discussing in some detail.

CHAPTER 3:
PANTHEISM AT THE CROSSROADS

IF THE TENDENCY TO BELIEVE IN A HIGHER POWER was the result of human brain evolution, one of the first influences on faith was undoubtedly the most basic socio-biological imperative - the survival instinct. Although there are more than 25 billion neural connections in the brain, human experience is rather focal. The most fundamental purpose of cognition, behavior and emotion revolves around the will to live.

Humans often act in perplexing ways. We believe in a wide variety of creeds, but all actions serve the purpose of keeping our minds and bodies intact. The organs of the body perform varied functions, but ultimately have the central purpose of providing nurturance to the cells and maintaining homeostasis. As Freud suggested, those purposes also play out in the psychological domain. Beliefs and feelings can vary, but only those that provide stability will prevail in the person's thought processand habit structure.

Organs provide for physiological stability. Values and attitudes provide stability for the self-system and the culture. All aspects of culture, including art, music, politics and faith exist to meet needs and provide stability. For that, and various other reasons it seems likely that the first religions were pantheistic.

Pantheism is rather interesting, because it is both ancient and modern, both primitive and theosophically sophisticated. It was employed in the earliest human tribes, possibly as the result of cortical brain expansion, and was passed down to Enlightenment philosophers such as Voltaire, Rousseau, and Nietzsche. Eventually it branched off into what is now called the Humanist movement.

Pantheism is exactly what it sounds like, assuming one has some familiarity with the Greek language. It is a 'wholeness' derived from the prefix "pan." In this system, nature is God and God is nature. As part of nature man is also included in the pantheistic paradigm. Pantheism assumes there is no specific authority, and that everything has the capacity to influence everything else. It is holistic, spiritual, and empirical, both religious and in some sense, scientific, because nature can be measured and, after all, nature equates with God in this system. Interestingly, the breadth of pantheism even extends to quantum physics with respect to the anthropic principle.

As a result of Werner Heisenberg's discovery of the uncertainty principle, physicists tried for years to explain why it was impossible to determine both the position and momentum of quantum particles (photons) and why, in various experiments, the act of measuring seemed to change the path of the particles, making them adhere to a more predictable, measurable type of motion.

It was as though the scientists were bobbing for apples, dipping down and attempting to take hold, only to discover that a wave-like entity moved the apple from their grasp. The eventual explanation came to be known as the anthropic principle. It suggested that since, in measuring the path of the particle, the eyes of the researcher emitted photons - as did the measurement device, the act of measuring was affecting the movement of the particles. That meant neither the researcher nor the instrument could be considered separate from the item being observed and that every aspect of nature was interconnected or "entangled."

That implied that the scientific method, the empirical philosophy, and the very idea that nature can be studied objectively could be called into question. According to this very modern concept, man can no longer be viewed as an observer. Rather, he is a participant in the process. Not a scientist per se, merely a co-active variable within the natural world.

The anthropic principle proposes that the universe is holistic, which coincides with the modern version of pantheism. It not only speaks to the breadth of this doctrine but also to its capacity to erase the long-held distinction between faith and science. It was especially influential in ancient times when there was no scientific method.

Despite having few modern adherents, pantheism has remained resilient, in part because scientists have come to consider possible unification within nature. That has made pantheism less subjective and uncomfortably mystical. Another reason for the staying power of pantheism is due to advent of the environmentalist and feminist movements. With that in mind, it might be interesting to look at pantheism in terms of how it derives from the cognitive development of our species.

It was discussed previously that brain expansion contributed to the development of internal language and the enhancement of attention faculties via the proliferation of frontal cortical cells. The evolutionary path of this expansion is uncertain, which means it is impossible to ascertain which hominids first began to employ such cognitive skills.

Each successive group of hominids displayed increasing brain volume. Homo habilis' brain was roughly 500 centimeters, Homo erectus' brain was 900 centimeters, Homo Heidelbergensis' brain was in the range of 1200 centimeters, as compared with 1500 - 2000 centimeters for modern humans. Since the skull case of hominids was bun shaped (occipital) and less frontal than humans, and since fossil records point to stagnation and repetition in tool making prior to Homo sapiens' arrival on the scene, it is possible the critical threshold needed to produce imaginative thought and attributive, categorical language had not been reached until the first archaic versions of sapiens appeared on the scene. However, it seems possible Neanderthal had that capacity. In addition, environmental pressures faced by prolific travelers like erectus and ergaster would have required a cognitive means of resolving conflict and summoning hopeful anticipation. Once that threshold was reached, it would have been possible to conjure up a variety of gods and belief systems.

The first versions of faith, and the first spiritual forms were likely based on need provision. Since the survival of nomadic groups depended on nature's resources it stands to reason that paying homage to providers of food, water, hunting opportunities, health, clement weather, and other factors made sense. In terms of human cognition, the reaction to obtaining these resources would have been emotional. Joy would result from enhanced survival opportunities. It was likely an era in which any separation between nature and God was inconceivable.

At the time pantheism arose, Homo sapiens was probably humbler. In nomadic groups he did not engage in conquest. He had no sovereignty to defend and no reason to boast about his riches, his imperialistic status, or his victories in battle, unless his group was threatened by wandering rivals. Nor did he have to worry about his legacy in a historical context. In effect, there was no psychological point in harboring excessive pride.

It is possible this dynamic was reflected in ancient cave paintings in Lascaux, France and Altamira, Spain. Though drawn 20,000 to 35,00 years ago, those cave paintings were quite detailed; not an easy task, since all artists had at their disposal were charcoal compounds. No doubt the drawings depicted something about the history and experiences of cave dwellers. Yet there are few if any drawings of humans, despite the highly social nature of archaic human groups.

Why no emphasis on humans? It is possibly because concern with survival correlated to such a degree with the outer workings of nature that the idea of the "self" had seldom been contemplated or necessary. Certainly, cave dwellers had ways of identifying one another, but it appears that did not matter much to them. In their functional society, concern over resources, hunting opportunities, weather patterns, the behavior of predators and the rearing of children would have been so urgent as to ameliorate the importance of individual identity. Only later, when urban, agricultural settlements allowed for down time was Homo sapiens able to reflect on his identity and the details of interpersonal experience.

While pantheism likely had its roots in life's realities for early man, its extension into modern times has featured interesting modifications. For example, during the Enlightenment period in Europe it evolved into a kind of compromise between Christianity and Agnosticism. Pantheism, then and now, has, as one of its central themes, that there is no transcendent God existing in a cloistered domain, i.e. heaven. There is only nature. All that matters are the territorial blessings found here on earth.

Pantheism existed within the Roman Empire but was forbidden by rulers after Constantine essentially made Rome a Christian republic. That was ironic because Rome had persecuted Christians in earlier times to enforce pagan worship. Then it carried over to Medieval times and prevailed during the Renaissance despite its apparent incompatibility with Judaism, Christianity, Hinduism, and Buddhism.

Yet, there is a point of contention to raise regarding the difference between pantheism and other faiths. It has been argued here that faith is dualistic - part God-derived and part man-derived. There is nothing new in the assertion that all religious systems have been influenced by evolution-driven functions of the human brain. For that reason, there is probably a point where pantheism is compatible with other faiths. For example, while pantheists did not believe in heaven per se, neither, to a large extent did the ancient Hebrews.

Judaism includes a distinctly spiritual outlook, but its description of the afterlife is somewhat vague. While the Jewish faith refers to Shamayin and Shoel as heavenly analogies, the afterlife was not mentioned at all in the Torah, perhaps because earthly concerns were paramount for a group trying desperately to preserve life, limb, and culture. Instead, the intent was to create specific laws and morals to utilize there on earth. In contrast, pantheists, were, and are primarily guided by self-direction rather thandeistic authority.

Both pantheists and Jews believed in personal responsibility and free will. They held the individual accountable for his actions. Most Jewish sects in ancient times believed the ideal reward for good acts was not Elysian Fields, as with the Greek pagans, or Lily Lake as with the ancient Egyptians, but a land of milk and honey right there on earth. References to heaven, particularly in Shamayin, focused on a state of reduced suffering rather than the pursuit of paradise. When all is said and done, both groups believed nature was the primary source of reward.

All religions have placed high value on the power of nature. Pantheists worshiped trees and water while Hebrews found their moral inspiration in mountains, burning bushes and parted seas. A similar comparison can be made between pantheism, Buddhism, and the Hindu faith. All religions seem to contain elements of pantheism, which one would expect from a human mind that evolved and adapted to a variety of often dangerous environmental settings. Consequently, the cognitive tie between nature's provisions and deistic regulation would appear to be unbreakable. As a result, one can assume this tie-in results from the structure and functions of mind, which of course, was shaped by nature.

Still, to presume that pantheism features a primarily omni-natural doctrine might be an oversimplification. It has been flexible enough to assimilate elements of many faiths. Pantheism has been practiced by Native Americans, Greeks, Romans, Confucianists and Buddhists. It had followers like Wordsworth,

Thoreau, and Einstein. Despite the various components of this broad belief system certain core elements are characteristic of the doctrine.

Some of these have been framed in notable quotes over the years. For example, Friedrich Nietzsche wrote: "All things are linked, entwined in love with one another." Einstein, who disagreed with Danish physicist Neils Bohr over the uncertainty principle and other elements of quantum physics, including the contention that matter could interact over "spooky distances" and that the universe was entangled nonetheless, wrote: "The individual feels the sublimity and marvelous order which reveal themselves both in nature and in the world of thought. He wants to experience the universe as a single, significant whole."

Philosopher Jean Jacques Rousseau also weighed in on the subject. He wrote: "I feel an indescribable ecstasy and delirium in melting, as it were, into the system of beings, in identifying myself with the whole of nature." Meanwhile, Wilhelm Georg Hegel wrote: "Reason (God) is substance and infinite power; its own infinite material underlying all the natural and spiritual life."

The core elements of pantheism become clear from these quotes. The idea that all is one and one is all... captures the essence of the pantheist theosophy. As discussed earlier, such elements of thought coincide with the evolutionary task of adapting to the natural environment.

Adding to its allure is that pantheism can, by virtue of its humble premise, provide relief from duress, which is the purpose of many religions and a prime motivational aspect of the human brain. We all suffer at times, and a belief system based on acceptance and submission to the natural world coupled with lack of egocentrism can ameliorate emotional burdens that are often a byproduct of excessive pride.

The idea that we are not special, merely a small piece of the environmental puzzle is integral to pantheism and can potentially lead to a life typified by peace and acceptance.

The problem with pantheism is that it reflects only part of man's survival pressures. Being so general, it tends to bypass the specifics of experience, as well as some evolutionary aspects of human development that have roots in primate social behavior patterns.

Among these tendencies are proclivities for social rank, competitiveness, inter-group and intra-group aggression, sexual opportunism, and the quest for material access. Feeling at one with nature does not alleviate the anguish of being oppressed and brutalized by foreign armies, or remove the anxiety and suffering resulting from illness, pain, suffering and the deprivation of love and social contact. In those instances, nature arguably becomes the enemy and destroyer of entire communities. Few people living in Paris, France felt at one with nature during the onset of the Bubonic plague.

Thus, pantheism left a lot out, and over time, did not address enough to address the needs of humanity, particularly once agricultural settlements became so densely populated that tribes began to compete with one another for land and resources.

At that point something more was needed; not that the need to commune with nature was tossed out the window of history. It's just that the human brain evolved into a supra-instinctive organ designed to deal with emerging contingencies through the learning process rather than being restricted by fixed patterns of behavior. The mind is pliable, able to change its emotional, motivational, and cognitive engrams through experience. Consequently, as the world changed, so did many aspects of religion.

Pantheism has lasted, albeit only marginally. It is certainly idealistic, hopeful and in many ways, admirably simple. It also provides a convenient bridge between faith and science, making the two bodies of thought less distinct than one might imagine. In fact, should scientists reach an impasse in search of ultimate answers, they might well end up hitching their wagon to a pantheist philosophy to gain closure. It would be an ironic conclusion to the quest for a unified theory but one that is theoretically possible.

Despite its breadth, pantheism has not been social enough and the human animal is a social creature. Because faith develops according to how we think and feel, religion had to become more social and complex. As populations in the first cities exploded, pantheistic gods evolved into more specific entities. The idea of God changed from a general, natural entity with no specific/limiting identity to a cluster of gods resembling falcons, bulls, and crocodiles.

So-called theri-anthropic gods were portrayed in combined human and animal forms: for example, a bull's head with a female's body, a crocodile body with a human head. A bridge was being constructed by changing times and human need. During this epoch, the forms assigned to God were, in some sense, a reflection of the needs, tribulations and conceptions of man. Finally, the stage was set. The man/God religious concept (a bridge leading from asocial to social deism) was becoming the new wave.

Three questions come to mind regarding that spirituo-political transformation. The ancients must have asked themselves, first, was the creator man or God? Second, is it possible to separate the controller from the controlled? Finally, if there is indeed reciprocity between God and man via the new hybrid model, was it possible God is so indigenous to human nature that, regardless of the deistic configuration, the mutuality implied in that relationship could never be undone?

CHAPTER 4:
THE PAGAN COMPROMISE

As DISCUSSED ABOVE, THE ADVENT OF RELIGIOUS SYSTEMS seems to have reflected human need and the ways in which human society is organized. That is true with respect to the transition from pantheism to paganism. The structure and purposes of human society have always been somewhat paradoxical, that is, both stagnant and progressive and invariably subject to the opposing influences of modernity and antiquity.

For example, recent human societies have espoused equality. Indeed, concern for equity in the population in those nations is now so fervent that despite variances in knowledge, social interest and loyalty among citizens, those societies have put the people in charge through free and fair voting rights.

The modern belief in equality is so intense that any hint of discrimination (even if only verbal or covert) is considered an act worthy of job termination, public scorn and even prosecution.

The emphasis on "power to the people" originated around 800 years ago with the signing of the Magna Carta. It was a milestone event. Some say it was a long overdue concession to human nature, others say equality - or populism - resulted from the threat of revolution. The question of whether true equality is consonant with human nature has been discussed by politicians, moralists, and philosophers, but has never been resolved; in part because anthropologists and psychologists, who understand human nature seldom engage in political discourse.

Another socio-political concept goes back further than the Magna Carta. Though ancient in origin, this version has existed within every facet of human experience from ancient to modern times: for example, in sports, politics, entertainment, literature, medicine, art, music, in the military and even among young children interacting on a playground. It is the concept of the social hierarchy.

Humans, and virtually all social animals operate by social rank. As unpalatable as it might seem to devotees of western democracy (bearing in mind that 'equality' initially referred to equal treatment under the law, rather than in terms of ability) we absolutely require a system of social rank to carry out our business.

The question could be posed as to whether equality or rank are native or learned human qualities. Some might say we must be "taught" to obey those in power through a process of Indoctrination, that it is natural to oppose authority and opt for equality.

Yet the evidence seems to suggest otherwise. One look at young children interacting makes it appear social rank comes naturally to our species. There are leaders and followers on the playground, in the military ranks, in school systems and vocational settings. Even kindergarteners will eventually organize themselves in some sort of pecking order.

There is conflict inherent in the hierarchical process, even among our closest genetic relatives, the chimpanzees. While they operate through social rank there are constant attempts by lower ranked members to overthrow the alpha male. The question is: Does this support the idea that social rank comes naturally to our species?

It is difficult to say. On one hand, chimps do challenge the rank order, but on the other hand, rebels seeking to depose the alpha male do so for purposes of elevating their own status. That suggests rank is so enticing that it is natural to primates. It seems, rather than obviating rank, those "primate rebels" seek to attain it.

In human society the same process prevails. Famous people are typically elevated, then diminished by the public. Fame can lead to perks and idolatry, but also to innuendo, scandal and even assassination. Once again, it is seldom the case that resistance against domination is pure. Instead by diminishing the human "alpha" the rebel elevates his own status. Lee Harvey Oswald killed John Kennedy to enhance his status, not to protest the presence of a leader. In that sense, the trend toward a social hierarchy appears to be ingrained trait, making it natural to our species. Still, it is hard to say why this is true.

It could be that as the social intelligence of a species increases it becomes easier to distinguish between and among members. That makes it possible to determine which members of a group are worthy of deference. Much of this is performance-based. The young student will look up to another because he or she proves more capable through physical or academic prowess or has an appealing sense of humor.

Is there a distinct, biological reason for social rank? No specific gene pattern has been found as pertains to dominance and submission. Obviously, there are neurochemical reactions that accompany feelings of dominance; for instance, an uptake of nor-epinephrine in the bloodstream. But neurochemistry can be an effect as well as a cause. It will be elevated after defeat of a rival or following a sexual encounter, but it can also be elevated as a preparatory mechanism to deal with the threat of being dominated.

In that sense, it appears dominance and social rank are the result of some combination of experience, neurochemistry, and inborn social traits. Those are powerful influences, which explains why hierarchical groupings are so prevalent in human society. As discussed previously, Pantheism, which entails belief in equality, humility and submission to nature does not take this into account.

Religion often brings out the best in mankind by creating lofty moral standards. It also allows for the opposite, by ignoring certain aspects of human nature that need to be controlled. That might be why, with the advent of agricultural settlements and increasing populations, the religious model had to change.

A gathering of large populations in one place can result in two possible outcomes. One is increased production because more workers can plant more, reap more, build more. To the extent the population does not exceed production and available resources that can prove advantageous. The other outcome is duress resulting from alienation.

Virtually all species tend to behave more favorably toward kin than toward strangers. The need to sustain genetic solidarity is fervent, and likely trickles down from genetics to social behavior. It is reflected, at least indirectly in the Old Testament. For example, in Genesis, entire chapters are devoted to specifying family lines and the ages of individuals - particularly patriarchs.

Much of the Old Testament was written at a time when agricultural settlements were densely populated, when tribes and family lines were mixing and competing. The threat of genetic dilution was palpable, which is perhaps why many of the practices were geared toward averting extra-genetic/cultural influence, including foreign ideas, faiths, and values. It led to cultural and tribal/familial compensations.

The Hebrews wrote with a passion about tribal and religious identity. The Egyptians sanctioned marriage between siblings to purify genetic lines and regal succession.

Faced with the threat of cultural and tribal dilution the drive for group distinction became intense. Emerging great civilizations assigned gods and beliefs to every conceivable aspect of life, which led to religious and cultural diversification. This was seen, not just in the Roman adoption of Greek deities but even in the first civilizations, which borrowed liberally from one another.

In Egyptian culture there were over 2,000 gods, many with overlapping functions and domains. Most were initially conceived in pantheistic fashion. For example, Aker guarded the horizon, and like pantheist deities was not personified. Ammut judged the deceased in the underworld, and had the head of a crocodile, body of a leopard and hindquarters of a hippo – an omni-natural conglomerate that seemed to blend with nature.

Egyptian gods were initially believed to be entrenched in the natural world, particularly primary gods such as Amun-Ra (God of the sun and air), Baal, the storm god (who was "borrowed" from the Canaanites) and Horus, an avian god who was very high in the deistic hierarchy. However, these God-concepts were not indigenous to a specific group. Rather, there was a blend of religions, languages and cultures which was subject to challenge.

One suspects the initial, and most pressing concerns of these fledgling urban societies was procurement of resources. In accord with the need/provider God-dynamic discussed earlier, access to food, water, ample sun for crops, and safe travel along the Nile would have led to expressions of gratitude. Given the neural machinery behind cause-effect thinking, it was inevitable that people in these settlements would extend their gratitude to powerful entities deemed responsible for providing those resources – entities beyond themselves.

One could argue that urban/pagan deities were the byproduct of a cognitive adaptation, combining nature-based pantheism with the increasingly material and personal concerns of agrarian settlers.

It is important to bear in mind that, despite the common view that agriculture represented an improvement in the lives of our ancestors, that is not entirely true.

Nomadic tribals had fewer mouths to feed. Most anthropologists (perhaps best exemplified by Dunbar's "natural group number") believe the typical early human tribe had at most a few hundred members. The math worked for them. Close knit groups tend to be more loyal, which favors child rearing, increases cooperation among workers, and lessens the tension between leaders and followers. Also, since nomadic tribal culture was transient, work efforts had to be renewed in each setting. That means the contributions of every member were critically important. That, in turn meant the value of each individual member would be maximized.

Ironically, the only real egalitarian human society might have existed in these nomadic societies. That does not mean early humans were devoid of social rank. It does suggest the practical tasks of work, the distribution of talents, divisions of labor and general concern with protection of offspring would have made "alpha-domination" less functional.

Once urban settlements expanded (the town of Memphis in Egypt grew rapidly to over 30,000 inhabitants) things changed. When populations increase, problems of a social nature arise. Cultural and religious mixing was only one source of stress. Having more mouths to feed, more group mingling, a pressing need to finance building projects, more intense competition for mates, and inevitable differentials in wealth resulting from the economic competition also produced enormous tension. Some individual or authoritative group had to modulate that tension to prevent social chaos.

An interesting irony was involved in this. Although leaders emerged as controlling agents, they could not decide on crucial matters without the support of aids, supporters, and armies. Despite their power, they never had numbers on their side. Over time, they quickly learned how to gather support through alliances. That gave them virtually absolute power.

They undoubtedly sought, first and foremost, to provide for their people, but as all leaders eventually discover, things don't always work out to everyone's satisfaction. More people in power (allies) were needed to rein in the frustrations of people who got lost in the shuffle and perhaps had distant memories of nomadic life when hierarchies were less necessary. In a small group, a decision by a leader would have a broad effect. More importantly, they could be observed and scrutinized by everyone. (With a little imagination one could draw a close comparison between this tribal dynamic and Thomas Jefferson's concept of "first grade purity" representative government.

However, with a massive increase in group size regal decisions might not meet with approval from all groups - especially when taxes were increased to pay for building projects most commoners could not see, enjoy, and perhaps did not even know existed.

Enter the doling process and advent of the class system. The haves had power the have-nots had none. The problem that subsequently faced all societies at one time or another cropped up. Specifically, that the poor had numbers on their side, and a vast potential army. It is an interesting feature of history that while powerful leaders typically mate more frequently, with more partners, the masses will tend to out-reproduce the rich.

As a result, there will always be more commoners than aristocrats. To the extent that such a disproportion might foment a violent revolution the initial urban leaders had no choice but to create laws that did not derive strictly from regal arbitration or sheer dominance. The laws could be class-tinged but had to be enforced much the same way for all people. Hammurabi of Sumeria was among the first to realize this. The Egyptians followed suit. Indeed, the need for laws to modulate social tension was a prime topic in the Book of Lamentations.

Hammurabi's code was an interesting example of post-urban human reasoning. It was almost absurdly proportional. For example, if a man injured his wife he would be injured as a penalty. While women were not considered as valuable as men, laws protected them in proportion to the weight of either their own transgressions or offenses against them. The laws were barbaric by modern standards but did create a future model for law and religion - influencing not only the biblical tenet... an eye for an eye, a tooth for a tooth, but laws within the English and American constitutions prohibiting the use of cruel and unusual punishment. Proportion became a new ideal in the management of complex human society, and for the most part it served the new settlers well. Over time, it would lead to a fusion of law and faith, particularly with Hebrews. While the need for apportionment was due in part to socio-political tensions, it derived, of course, from the human mind.

What is it about the human brain that fosters proportionate thinking? Proportional thinking appears to be a built-in feature of the human brain, and while it needs to be refined through learning, the neural template is present at birth. Even children veer toward this cognitive scheme.

As an illustration: a little fellow wants to take his motorbike out on a slippery road. The parent refuses to allow it. The young fellow protests, saying (perhaps) *Johnny down the street was allowed to take his bike out - why can't I?* The parent, then issues a proportionate syllogism of her own, to wit; *If Johnny jumped out of a plane without a parachute, would you do that too?*

In both instances, a cognitive balancing act takes place. It is an equation of sorts, whereby one person's behavior is weighed in importance and meaning with that of another. This is course, a staple of not only cognitive reasoning. It is the foundation of mathematics as well. The fact that it can encompass everything from the statement of the little boy with the bike to Isaac Newton's invention of calculus suggests it is an essential human cognitive mechanism.

The question of how this neuro-psychological skill developed is interesting. In the transition from occipital (hindbrain) primacy to frontoparietal expansion, the brain lost some degree of functional specificity. The occiput, which was crucially important for visually oriented primates, has specific pathways from the retina to the optic nerve and to the primary cortical association areas. Conversely, as neural clusters expanded toward the frontal cortex (giving humans the signature highbrow look) the functions became less specific. The frontal lobes do not have a specific sensory, motor or mnemonic purpose, which is why removal of frontal tissue does not lead to significant cognitive impairment.

However, as discussed earlier, the frontal lobe does have enormous influence on brain/behavioral functions. Due to its functional ambiguity, yet inter-connective influence, the frontal lobes monitor and regulate the activities of other brain sites. To perform that function the frontal cortex must be capable of extraordinary parsing. The reason is that, despite its reputation as the circuit devoted to things like attention, self-regulation and planning ability, the frontal cortex has connections to more parts of the brain than any other circuit. Those vast connections can produce significant amounts of noise. Only a sophisticated computer with superb parsing capacities could wade through such turbulent waters and reach conclusions. To do that requires an ability to hold some inputs in abeyance while allowing others to pass through. In other words, it must have exceptional capacities for selective inhibition and excitation - key neurological components of proportionate thought.

In effect, the frontal drift in evolution created a kind of electro-chemical circuit breaker. Because of its influence in the brain as per its vast connections and regulatory purpose, the frontal cortex was selected by nature to serve as a gateway to reason, morality, mathematics, and consciousness itself.

A capacity to apportion thought was not necessarily extant when the human forebrain expanded. Many of the skills and technological advancements developed and achieved by our species could only be manifest once sedentary societies created down time, which freed up the imagination.

In that context, many paleoanthropologists have asked why human culture has been so punctuated. For example, the modern human brain has been around for 250,000 years, and even if one assumes the most recent version of Homo sapiens emerged 50,00 years ago, that is still an enormous amount of time for man to wait to create cultural technologies. The question of why, rather suddenly (in evolutionary time scales) man went from cave-dwelling to the construction of pyramids has been pondered many times in scientific circles.

There are probably two reasons for that delay. One has to do with time and opportunity. The other with what one might call population-induced generativity. Nomads do not have much time to dream and create. However, once the permanent settlers had down time there was a rapid improvement in building methods, medical breakthroughs, agricultural innovations, literary styles, political theories, and religious principles. With more time on his hands, man's potential came to the fore. Leisure time allowed for interpersonal discussion of ideas and an unleashing of the imagination. Each new idea branched off into another.

The generativity factor is complex but one apparent cause of a rapid increase in creativity is competition. Living in highly populated settlements meant large numbers of people could observe one another, see what others have accomplished. They could copy methods, try to out-do their competitors as well as their comrades. Much of this, incidentally, was driven by the natural tendency to enhance one's position in the social hierarchy.

Beyond competition, there was a mechanism one could call the 'rational stairway.' Once an idea or invention has been produced, the associative neurons of the brain will tend to expand on that. As a musical example, in the 50s, basic rhythm and blues evolved into rock and roll, as musicians took the rudimentary A-B-A format and set it to a faster tempo. Rock and Roll was simply a variation on a theme.

Another example is seen in literature, which expanded from traditional storytelling, involving fictitious plot and character, into a more journalistic style as exemplified by Truman Capote and Norman Mailer. In a broader context, human culture and technology have always advanced because of the gift of collective memory and the quest for social rank. Population density contributed to that in the earliest forays into creativity.

In essence, the newer ideas and inventions could not exist without knowledge and modification of previous ideas because learning begets learning and change is a cognitive mandate. Homo sapiens is, and has always been the ultimate artist, as well as innovator who is both irritated by and attracted to uncertainty and novelty.

That process was perhaps best captured in a quote from Sir Francis Bacon in his book Advancement of Learning. He stated: "If a man begins with certainty he will end with doubt, but if he begins with doubt he will end with certainty."

The implication of this statement (which is a concept developed much earlier by Socrates via the dialectic teaching method) is that the human mind is most essentially an interrogatory machine. It evolved by adapting to varied environments, particularly as hominids like Homo erectus and Homo ergaster adopted long distance migratory cultures. Faced with the task of coping with newness on the fly, regarding terrain, resources and predatory threats, early humans faced challenges. Fortunately, two advantageous traits helped them adjust: curiosity and anticipation.

It had begun with the upright posture, which created the capacity to 'look ahead.' That rudimentary trait, combined with a voluminous noise-heavy brain favored humans with proclivities for closure-seeking and anticipatory cognition.

Nature works most of her miracles through the process of pleasure enhancement. For a trait or behavior pattern to persist, its execution must produce either pleasurable feedback or relief. The pleasure need not be appetitive. It can simply be found in resolution, whereby the noise and confusion resulting from activation of millions of neurons is resolved after closure is attained. Psycho-physiologist Daniel Berlyne wrote about the pleasure/closure process which he referred to as the arousal jag. There is a parallel between these brain dynamics and the transition from pantheism to paganism.

The transition was probably not fluid, and undoubtedly occurred in fits and starts. But, once again, the mind of man did the orchestrating. The logic and apportioning mechanism took over.

Having expressed gratitude to nature for bounty and survival, early tribal nomads could comfortably pay homage to nature in a direct sense. All they needed was game, rain, the sun, the sky, the earth, the waters, and the trees.

As towns cropped up, life was becoming more secular and political. At some point, kings (who were initially treated with great skepticism), gained increasing control, and had expanding influence over the average person. The king became powerful and the prime giver of wealth, work, and civic access. The same traits previously reserved for a natural deity were attributed to monarchs. There were still gods of nature in every society. In fact, though the power of pharaoh increased in Egypt, the number of gods increased, largely because for some reason, when Egyptians changed preferences for one god over another, they seldom abandoned the previous god.

Despite being retained, many gods were diminished in status. One God could be responsible for ensuring safe travel for boats across the Nile or escorting the dead to the hereafter. Another could be responsible for maize production or fertility. Yet, as the deistic division of labor expanded in concert with the increased complexity of agrarian society, each God became sequestered, unable to cross into another's territory, or, for that matter serve the people outside their areas of specialization. Over time, and with the increasing politicization of agricultural settlements, the king was able to compete with God.

That was in part because God intervened in actual events only rarely and did so through the mechanisms of nature. Faced with a drought and diminished crop yield, a farmer would have to wait for rain (for God knows how long). If a wife could not get pregnant, it might take some time and endless prayer for a fertility god to come to the rescue.

Kings, on the other hand, could solve problems faster, and since many problems originally deriving from nature were becoming more civic and gods more specialized, the king could address any number of issues and concerns as a singular controlling force. That might explain why there was an inevitable merger between king and God.

The man/God model cut through the red tape. Still, man-gods lacked a certain something. Men can be ruthless after obtaining power. Lincoln once said: "absolute power corrupts absolutely" and as monarchs began to compete with gods - or in many instances declare themselves gods, that became painfully clear.

Amidst this trend, conflict ensued. Answers were needed. While the pantheistic deities viewed people not only as equal but no more important than a bull or a falcon, the kings began to differentiate among types and classes. Over time, there were loyalists and rivals, poor and rich, young, and old (particularly as pertained to their capacity for work) and finally, citizens and slaves. It was ironic. Once trapped within the confines of urban settlements, the average person came to rely on the provisions of the state. As Plato wrote, the state made man, gave him culture, sustenance, and an identity. However, to sustain the state meant having an army and the army was controlled by the king. That gave the king potentially absolute power. The same resources and controls enabling the king to defend the sovereignty could also be used to repel uprisings and suppress complaints. In effect, nationalism began to replace pantheism in this new sociopolitical model.

Given the high level of tension resulting from this transition, the people needed solutions. Interestingly, the next religious transition was founded as much on the class struggle as on spiritual concerns.

The first solution took the form of paganism. This religious format is similar to pantheism in that it is difficult to describe it as a step in religious development. The main difference is that paganism personifies the gods, even to the point where they have human-like foibles.

With the advancement of this belief system there was still a division of labor. In Rome, Mars was the god of war. In Greece, Dionysus was the god of fertility. However, these gods were treated as genuine decision makers. They could punish or reward. They had an agenda. No matter how formidable an army, its leaders would typically visit a temple and consult with and ask for guidance from those deities.

Pagan gods were not described physically for the most part - though many were assigned physical traits such as strength and wisdom. It represented a cognitive drift that, in a sense, reinstated a higher authority into the cultural equation. That provided restraint on the monarch's power. Was it enough to assuage the masses? In some instances, yes, in others, no. If the king was fair-minded, as Hammurabi tried to be, things might work out. If not, an appeal to the gods by a slave might go unanswered.

Since there weren't many leaders as wise as Hammurabi, the lower class began to look upward for entities to address their problems. However, this ran into a snag. Many of these peoples were used to a pagan system, and even if they had prior familiarity with pantheism, there was a problem. Pantheism is not so much a problem-solving system as one of attribution. In the aftermath of a successful hunt the faithful could pay homage to the God of nature. However, to provide a solution for an existential problem was somewhat beyond the scope of a pantheistic God. Pantheism is heavily weighed down by the need for blend and acceptance. It does not coincide with states of irritation and unrest, and in that sense, is not often the religion of choice for the poor.

A revision was needed and required the emergence of leaders. Some were prophets, others hard core revolutionaries. And one more thing was needed. If oppressed people were fortunate enough to find God through a prophet's insight, the god would have to be capable of communicating with and commanding generals, kings, and other powerful people in control of various resources.

To create such person-to-person impact, the new Gods had to be conceptualized singularly. This version of God could not be merely a natural force granting bounty to travelers. Instead, he or she had to govern all aspects of life. In accord with such criteria, a new model cropped up. It challenged all prior religious models, not in the sense of refuting ideas inherent in those models, but rather by amalgamating those ideals under the command of a single, all-powerful entity. It represented a move toward passion and convenience, and, as always, need was coupled with faith. The new format, monotheism was both controversial and seductive.

CHAPTER 5:
AKHENATEN AND ABRAHAM

THE ADVENT OF MONOTHEISM was, in one sense, a new phenomenon. However, while it first appeared around the second millennium B.C. it was arguably an inevitable extension of older ideas, and perhaps endpoint of a bridge between one mindset and another. Its exact timelines are difficult to ascertain, not just because record keeping skills were primitive, but because like most other cultural trends involving a transition in beliefs, there are usually overlapping times and locations. Had human society back then been dominated by leviathans like Rome or England, whose cultural influence spread over a wide swath of territory, our understanding of its origins might be clearer.

It seems one starting point was in Egypt, which was one of several powerful national/tribal entities through which travelers passed, along with their ideas and customs. Other such central destinations were Sumeria and a patch of territory occupied by the Hyksos. Among the many travelers passing through was a group who appear to have migrated from the Akaad. They would eventually refer to themselves as Hebrews.

Many historians believe there were two main authors of monotheism: the Egyptian King Akhenaten and the Hebrew-Aramean Abraham. Each lived in highly populated urban areas. Both left their places of origin in search of monotheism-friendly environments. Each had somewhat different reasons for their conversions and to underscore their religious transformations both changed their names.

King Akhenaten was born with the name Amenhotep. The Aramean who eventually referred to himself as Abraham was born Abram. The former changed his name to express his newfound allegiance to one God and to embellish his status as that god's main advocate. The god was Aten - the sun god, Thus King Amenhotep became Akhenaten. He also referred to himself in other ways. One version meant, "One Who is Effective for God." Another version signified "Son of God" – a familiar title that would travel through history, as both Alexander the Great and Jesus of Nazareth adopted the reference.

Just why Akhenaten decided to single out Aten is not really known but his belief was passionate. He established an entirely new city named Akhetaten to honor his god and while some historical accounts have him prohibiting worship of all prior deities, other historians believe he was more tolerant. In any event, Aten became the prime god during the reign of Akhenaten.

While it is difficult to discern the latter's motives, it does seem building temples to his god in a location removed from the central seat of power, coupled with his decision to demote, if not reject the existence of all other gods was very personal. This was not just the God of Egypt. It was Akhenaten's God of Egypt – his heartfelt revelation. Indeed, so intense was his devotion that he effectively rendered irrelevant the highly revered Sun God, Amon.

That gesture created controversy and Akhenaten paid dearly. However, the king was willing to undergo public scorn, proving this was not merely an ego trip. To the contrary, his devotion to Aten clearly reflected a strong belief in a higher power, not simply an exercise in self-aggrandizement.

The attempt to establish monotheism in Egypt lasted only as long as Akhenaten himself. His successor, Tutankhamen returned Egypt to the old pagan/polytheistic system after Akhenaten died. The question remains, however, as to why there was such a controversial turn to monotheism.

To speculate requires some discussion of the subject of attitude change, which involves specific psychological factors. Cognitive habits are hard to break. Indeed, researcher Leon Festinger devoted a good part of his career to study of a phenomenon called cognitive dissonance. He discovered that attitude change is often prompted by competing, discordant beliefs, and that duress can often be a prerequisite to a change in beliefs. For example, an alcoholic in treatment will often come to believe in a higher power to alleviate anxiety rising from, on one hand, a compulsive, anxiety-fueled need for control, and on the other, lack of control over a destructive habit. By surrendering to a higher power, his anxiety can be alleviated, and hopefully, so might be his dependence on alcohol.

What duress existed for Akhenaten? He was a king, therefore a man of privilege. Yet, in some sense he seems to have been conflicted, perhaps even desperate. Writers such as Mehaera Lorenz discussed some possibilities. One such possibility is that Akhenaten suffered from Marfan's Syndrome, a genetic disorder characterized by elongated features, extreme tallness, malformations of body tissues and strong sensitivity to cold temperatures. Was Akhenaten's devotion to Aten a plea to the provider of warmth to alleviate his discomfort, or perhaps a means of assuaging his embarrassment at looking odd?

Various accounts suggest Akhenaten was hidden from the public during much of his childhood. Also, he died at a relatively young age, and his brother Thutmose died even younger. That is significant, because Marfan's is genetically transmitted within families and typified by a shortened life span. Beyond that, pictorial depictions of Akhenaten and family members feature elongated features. If hiding was his default position it is revealing that his decision to move his holy city to a remote area and build a temple to Aten was both an act of faith and an escape-withdrawal tactic

There are several possible explanations for these tendencies. One is that Akhenaten had his depictions drawn in elongated form to symbolize his exalted status. Another revolves around his health status. Perhaps the most widely accepted is that he did have Marfan's Syndrome and that the depictions were accurate. After all, embellishing the height and wing spans of previous pharaohs was not common and the pharaohs probably all had substantial egos.

Still, another possibility is that he had become erratic, and eccentricity led him to commission unusual depictions to make him appear transcendent. For example, he often referred to himself as "The Unique voice of Aten." It is also conceivable that because his affliction made him self-conscious, he figured out a way to turn a weakness into a strength by suggesting his odd physical stature signified his uniqueness.

In any event, one can assume Akhenaten's decision toward monotheism was, driven in part by internal motives. He apparently needed to worship one God, not just due to his role as pharaoh, but for his own peace of mind.

There are commonalities within various versions of monotheism that contrast sharply with paganism. A plethora of gods, as seen in the pagan systems, offers no existential relief, especially if the deity rules over specific domains. Human beings in states of physical or psychological suffering need a single, reliable caretaker who knows them, can relate to them, and can provide the compassion and comfort typified by the old-fashioned roles of parent or caretaker. Pagan gods had power but did not typically connect with people. In some ways, their exalted status ruled out the possibility of compassion. Because of their parsed responsibilities, they had to be detached, objective and categorical in their concerns and actions. Theoretically, and in a somewhat comedic context, a plea to the god of the sea for a better harvest might result in an "it's above my pay grade" response.

Perhaps Akhenaten felt a need for comfort that only one God could provide. Had he been around in modern times he might have challenged the fanciful but questionable notion that: 'it takes a village to raise a child,' with the reply: "A village can watch over a child, but only a parent can love a child."

With respect to cognitive factors behind this, one can draw comparisons between God and the 'father' archetype inherent in many religions. Freud wrote about this in *Civilization and its Discontents*. It seems the typical family unit throughout human history has always included a single patriarch. There might have been more than one wife, but there was always only one father. I am aware of no society, ancient or modern, in which one woman was married to a group of men at the same time. The question is whether a cognitive template was handed down in brain evolution that fostered a dominant, single father stereotype.

It seems doubtful that humans are genetically programmed to view males as being superior to females. However, it does suggest that since our remote ancestors began as primates, our genetic-behavioral history could have ancient parallels with the male dominant dynamics of most primates (bonobos being an exception).

In some ways, this trend among primates is rooted in biological pragmatics. Since males can replenish their sperm count roughly every twenty minutes and females ovulate only monthly, males tend to be more sexually solicitous, have a greater capacity to control family structures and have more genetic influence. Once again, this is not absolute. Bonobo chimp sexual patterns are more egalitarian. It is also true that for primates, sexual activity is not just for reproduction. It is also a way to avert aggressive behavior, a means of relieving tension and a bonding activity. Nonetheless, the drift toward a patriarchal structure in faith might have been one factor leading to a decision by Akhenaten and others to gravitate toward monotheism.

Thus, it might have been inevitable that monotheistic beliefs would eventually reflect the social structures of our species. As Stephen Hawking has written, the human mind operates through established templates such that we invariably interpret experience in terms of human nature. While we can look outward to observe, gather data and draw conclusions about the physical universe, we have no choice but to perceive according to the sensory and cognitive engrams within our brains.

Once again, the man-inspired elements of faith might derive from neuro-biological traits with which we are endowed by nature. This process appears to have played a role in attempts by Akhenaten and Abraham to introduce a single, all-powerful god into the human equation. In the truest sense, this does not make religion less valid. Rather, it makes it more valid, more real and indeed more ingrained in the human mind.

Akhenaten is assumed to be one originator of monotheism, but he came on the scene many years after a Sumerian Semite named Abram. Akhenaten was born somewhere around 1370 B.C., Abram, around 1900 B.C. Like Akhenaten, Abram decided on a name change - this time at behest of his God. The change in moniker portended his future and that of three great religions. In Genesis 11 and 16 he is called "Abram" which means "Exalted Father." Then in 17:5 and 17;15 he is referred to as Abraham, which means "Father of Multitudes." That suggests not only that this modest man; a shepherd and donkey trader, was destined to be leader of a tribal family but that he would be the seed from which new nations and three religions would grow.

Interestingly, there are inconsistencies in accounts of Abraham's life. He is described as a mere shepherd whom God summoned to travel from his birthplace in Ur, to Egypt, Heron, Bethel, Canaan, and Jerusalem at various times in his life. As was depicted many times in the Old and New Testaments, shepherding did not provide a lucrative lifestyle. Yet, from the beginning, religious texts make it clear that Abraham had power and influence. God summoned him to travel about, but to do so he needed both social and material resources. That suggests he was more than a poor shepherd. To the contrary, he seemed to have considerable status. He was able to rescue his cousin Lot from imprisonment. He got involved in the political travails of Sodom and Gomorrah. He had enough wealth to pay a tithe to the king of a new township known as Salem (later called Jerusalem). He entertained angels, and he did business with royalty on behalf of his people in a variety of locations.

In some ways, his situation was similar to that of Jesus of Nazareth, who was the son of a carpenter from a small town whose family was so poor that he ended up being born in a manger, but on the other hand clearly had the theological sophistication of a scholar at a very young age. Nowhere in the New Testament does it discuss how he became so learned - bearing in mind that being literate was rare in his day, especially for a laborer.

It is possible the descriptions of both men were metaphorical – a descriptive tendency during the times in which both men lived. A shepherd could indeed have been a man who brought animals to a livery for trade or who profited from selling wool in a marketplace. Or it could have been used to symbolize a leader of men who led them to a new socio-moral realization.

While both Abraham and Akhenaten were men of influence, and probably the originators of monotheism, their circumstances were quite different. Akhenaten was an isolationist who diverted from mainstream Egyptian beliefs and sites of worship. He had been hidden during his childhood. He was unusual and iconoclastic, and perhaps, due to a need to compensate for his self-consciousness, was egocentric as well.

In contrast, Abraham was broad-minded and universal. While his conversion to monotheism resulted from direct interactions with his God, his relationship to the latter seems to have evolved into something more fraternal than paternal -

though it changed when God asked him to sacrifice his only son, Isaac. Abraham was also a cosmopolitan man, comfortable in whatever culture or town he found himself.

The irony inherent in a culturally eclectic man like Abraham opting for monotheism raises some interesting questions about his mission. He is not considered the father of a single religion but of three: Judaism. Christianity and Islam. His life was not urgent and predetermined like that of Jesus, who ended up fulfilling his sacrificial mission within a three-year period. He supposedly did not marry until the age of 85, sired his first child a year later. His wife Sarah could not bear him children, so his first child was born to her servant, Hagar, Abraham's first wife. Their son Ishmael became the father of the Islamic faith. Interestingly, his God did not have misgivings about this, indeed expressed pride by saying Ishmael would go on to become leader of a great nation. Only later, at God's behest, did Abraham and Sarah give birth to Isaac, when Abraham was 100.

This series of events says something about Abraham and the times in which he lived. He came from a pantheistic culture. Sumeria was the flip side of Egypt. While Akhenaten chose Aten as his prime deity, worship of the Sun God (Ra) was previously ingrained in Egyptian religious practices. In Sumeria the main pantheistic deity was the moon God, Sin. Why the difference, especially since, unlike the sun, the moon did nothing to enhance agriculture? It did not light up the day or nurture plant life to provide food for the dense populations in Egypt. Nor does it provide warmth. But while crops, the disposition of the Nile and the need to feed multitudes were constant concerns for Egyptians, the Sumerians were a bit more mystic. They were superb astronomers and mathematicians who used celestial phenomena to make predictions and fortify religious beliefs.

Abraham grew up in a milieu entrenched in magical beliefs. As mentioned in Joshua 24:2 his father, Terah was an idolater and pantheist/ polytheist (used interchangeably here because belief in gods of nature rather than personified beings is common to both) and apparently had no profound influence on Abraham's ultimate quest for a single god. However, being from the same culture, both father and son believed in an all-powerful god. While Sumerians worshiped many gods, they, like virtually everyone living in a polytheistic society, believed in a deistic hierarchy. The god in the loftiest position was deemed capable of overseeing all affairs of nature.

Once again, however, even an omnipotent god cannot meander into a shepherd's home, counsel him on whom to marry or dictate where to grow crops. Neither Baal, An Enki, Amon Ra, or any other prime deity could reach down to the general population to serve as house physician. Since the largest segment of the population in any town (then and now) tend to be of the lower and middle class there has always been a need for apopulist god.

One suspects, like Akhenaten, Abraham was really in search of a populist, personal deity - one who could be summoned, a god in touch with the times and the experience of his creations. Both men wanted a god they could take to heart, not just an entity to whom one offered animal sacrifices. In other words, both Akhenaten and Abraham seemed to be responding to duress resulting from an existential vacuum. Was this because crime and alienation were increasing in heavily populated towns, with tribal cohesion being threatened by modernity? If so, then one might assume both men were really trying to introduce a paradigm into the religious dynamic that was a throwback to nomadic times when dependency among all members in the tribe was strong, personal and, arguably bio-socially natural - a sociopolitical model that did not emphasize doctrines and did not utilize restraints and punishments to induce cooperation. Instead, it was natural and comfortably ingrained in the networks of the human mind.

In that sense, it appears that while monotheism marked a transformation of faith, it was merely a variation on the theme of basic human nature. The evolutionary programming that gave us social intelligence, empathy, a template of mind for what constitutes "the family" and the neuro-frontal capacity to facilitate self-observation, introspection, and an ability to compare our thoughts and feelings to that of others (often referred to as a Theory of Mind) finally exerted its influence on religion.

In some ways, this represented a deviation from all prior spiritual beliefs. It represented purity. After all, we are all creatures of nature. The sun and moon, the waters and the fields, the birds and fowl, mountains, and rivers – all have an impact on our lives. They are worthy of reverence. Yet rivers and mountains do not speak, cannot embrace, do not understand pain and suffering and have no capacity to offer hope in times of uncertainty. With increasing numbers of people clustered in densely populated towns, run by powers in remotely situated temples and castles, the thing increasingly alienated people needed - direct psychological contact was becoming at once more necessary and less attainable. Into that void stepped the gods of Akhenaten and Abraham.

Their new creed caught on. Later, the Sumerians leaned toward a monotheistic system. A group called Zoroastrians developed a deistic hierarchy that all but eliminated peripheral gods. Sin was their El, their Aten.

In truth, the idea of a purely monotheistic religion has never been fully realized. Akhenaten did not completely prohibit festivals and rituals devoted to other gods. Abraham believed his one god was all powerful but showed respect for other gods. In fact, his Hebrew descendants forbade the deprecation of other gods. In Exodus 22-28 Jews were warned against mocking the gods of other nations. Beyond that are references in the Old Testament to myriad "gods" - not "god" - despite El's primacy. For example, in Genesis 1:26 God says: Let us make man in our own image (emphasis on the word *our*).

Even in the annals of Christianity there were angels, saints and prophets with God-like powers and characteristics. Also, Isaac, Isaiah and David all had special relationship with God and could be said to comprise, at the very least, an upper echelon within an ostensibly monotheistic system. Also, despite fervent worship of Allah, the Muslim faith features a hierarchy of greater and lesser prophets.

That lends support to the contention here, that since the thoughts, feelings and conceptions of man can only arise from a brain designed by nature, we are relegated to patterns of worship as per our neuro-psychological wiring. It would seem the neural configurations of mind prevents us from conceiving of entities and groups outside the usual social, familial and relational parameters we are programmed to see.

However, times change, and within that neuro-experiential framework so do our ideas, needs and solutions. Akhenaten and Abraham were visionaries who, like Jesus, were able to read the signs of the times. As a result, their revisions set the stage for a new world order and a new mindset for the species Homo sapiens.

CHAPTER 6:
GOD, SPIRITS AND PEOPLE

While the monotheistic movement was underway, human attitudes mitigating against the idea of any type of single ruler were still in play. Because humans experience a level of discomfort when detached from social groups the idea of a community of gods couldn't have been discarded no matter how innovative the monotheistic format. While the archetype of "father" was important, so was the archetype of "mother" - and for that matter other members of the core family group. Man, after all, is a highly social neo-hominid, not just a patriarchal primate.

Perhaps, due to our social dispositions a group of mediators was inserted into the Abrahamic religions over time. These entities were variously referred to as angels, seraphim, cherubim, and messenger spirits, and they filled a void.

It was an interesting addition to the theological lattice, because while God the father was deemed omnipotent, omniscient, and omnipresent, it was decided he needed an army of formless soldiers, who did not die, did not reproduce, were not material and were a notch above human beings who ironically, God, decided (in Genesis 1:28) would rule the earth.

Why the need for intermediary figures? Reverting to the original argument here, it was conceivably because the demands of faith were urgent. No matter how pure their occasional consultations with a single, busy God, people required immediacy.

God has been conceptualized in various ways over time. Atheists have viewed him as a non-essential human invention. Christians, Jews, and Muslims see him as real, transcendent and a being whose acts are beyond human comprehension. Here, the argument is for a middle ground explanation that incorporates two elements. One is external and proactive. It includes a God of nature who governs all things, while providing a format from which the difference between moral and immoral acts can be discerned. The other is internal. It emanates from within the human mind - perhaps from a neuro-behaviorally determined tendency to look outward for explanations. The latter feature is consistent with what theologian Karen Armstrong wrote about human begins; that we are innately, unavoidably disposed to religious worship.

Then again, why the need for religious transcendence in modern times, especially since science continues to answer questions about our world? Perhaps, because despite its precision, scientific inquiry sometimes leads to an endless void.

For the past several decades scientists have searched for a theory of everything. They have not found it. While there are several versions of this, including super string theory, pilot wave theory, holograph theory, and M theory, (or membrane theory, as it is also known) all these ideas do is address the behavior of force and matter. Even if scientists reached the point where they could explain gravity in quantum terms (which would require a gravity particle, rather than a space-time warp model) they would still not have a theory of everything.

For example, how can unification of the strong, weak, gravitational, and electromagnetic forces describe why a sociopath can behave altruistically at times? What can force-matter interactions and the uncertain pathway of particles tell us about the need for identity, which can lead some to higher achievement and others to join a cult? What about the tendency of some to horde and others to share, or the process by which a loving couple descends into mutual antagonism and divorce?

Questions and dilemmas like that cannot be measured, no matter how many equations are written on a blackboard at M.I.T. or Cal-Tech. Religion, on the other hand, attempts to explain, and when possible, remediate such problems. Religion provides comfort when duress is overwhelming and uncontrollable through support and theosophical perspective.

While all behavior patterns ultimately derive from biochemistry, the volume and complexity of human behavior makes prediction impossible in scientific terms. Attempts have been made to make such predictions; most notably by adherents to what is loosely referred to as behavior theory. Legendary scientists such as B.F. Skinner and Ivan Pavlov attempted to explain all aspects of human behavior deterministically, but other than in very broad terms they were not successful.

In that context, perhaps one reason man veers toward the spiritual world is because it's the only way to answer certain questions that cannot be addressed in any other way. The fact that the human brain is a closure machine makes that quest unavoidable - which makes faith necessary.

Religion is often equated with spirituality, but that word has so many meanings and connotations that it's hard to describe, let alone put in a religious and evolutionary context. It is possible to attribute the capacity for spirituality to our neural software, because for the past 50,000 years humans have had that capacity. Why and how they used that capacity varied, but one constant feature seems to have been a belief in ethereal, mystical figures.

What is a spirit? Obviously not a flesh and bones biological entity. It has often been equated in modern and in ancient times with powerful emotions. A modern example of spiritualism is seen in the music of Carlos Santana. He not only considered his mode of expression spiritual. He felt somehow obligated to perform his music inthat context.

Listening to music can be exhilarating. Santana's guitar riffs certainly fit the bill in that respect. But what does even that mean? One answer might be that it conveys a sense of transcendence, which in this case means a feeling not attributable to a specific cause or antecedent - in other words, a naked but profoundly arousing experience.

If unexplained, exhilarating feelings comprise one element of spirituality. The 'unexplained' aspect is the key. When things happen that cause an intense emotional response that we cannot explain, it creates a kind of void within the central nervous system. One reason for the void is that the brain becomes pan-activated in searching for associative memories that lead to solutions, i.e. closure. Yet the human brain is not designed biologically or functionally to deal with voids. The vast human brain evolved to, among other things, solve problems. It was a necessary cognitive trait in response to the potential for "noise" resulting from the overwhelming interplay among billions of neuronal interactions.

One key to survival among all creatures is response selection. An effective brain must be able to clear its pathways to find a response apropos the circumstances in which the actor finds himself. Neural cluttering simply won't do in that respect.

The drive for closure is one reason human beings are capable of both logical and illogical beliefs and actions. Moralists and jurists often discuss the importance of ethics in terms of the functions of any given society, but the brain's primary purpose is to come up with interpretive closure so that responses can be selected with felicity and the general arousal that can reach intense proportions can be resolved.

In the above context, it is easy to see why spirituality is part of human experience. Regarding Carlo Santana's music: if he were to hypothetically impart to listeners the notes, time signature, chord progressions and other nuts and bolts aspects of his songs (in other words, objectify the otherwise inexplicable, subjective elements of the music) one suspects his music would seem less spiritual.

If that paradigm makes any sense, then one might expect the level of spirituality, including attributions, actions, and beliefs to have greater importance in a world of uncertainty.

Prior to development of the empirical philosophy, beginning with the work of Sir Francis Bacon, the advent of the scientific method through the work of Descartes and Galileo and the creation of the Arabic decimal system, the world was very uncertain. Still, the brains of these pre-empiricists were the same in volume, structure, and function as those of the post-empiricists.

Consequently, they needed closure. Spirits filled the void. The question of how, in neurological terms, nature gave us the capacity for spirituality is not difficult to answer. Much of it has to do with memory, which is itself a miraculous thing. Being able to conjure up images and experiences that are not occurring in the present is spiritual. Memory defies the laws of physics, it re-aligns the real with the imagined, treats the past as a workable present, enables us to catalog thousands of experiences, almost regardless of how much brain matter we lose over time. Even as we get older and our brain cells deteriorate, we can become wiser. In that sense, memory transcends biology as well.

Some theorists have attempted to explain that aspect of memory. For example, neuro-physiologist Karl Pribram attributed memory to a holographic process which enabled the brain to store information economically and redundantly, so that the sheer amount of brain matter was not a critical determinant of mnemonic efficiency. Meanwhile, after studying memory and concluding it existed in no specific location in the brain, renowned neurologist Karl Lashley considered the possibility that there is no strict neurological explanation for how memory works. In the final analysis, the mechanisms underlying human memory have yet to be determined.

It is certain that while God ostensibly made man in his own image, nature turned humans into a vast, virtually unlimited library such that a person from generation A can pass down his knowledge to a person from generation B, thereby making person B more knowledgeable. Memory creates a fabulous paradox by enabling those with less intelligence than the great inventors to accrue more knowledge than those inventors. (This writer knows more than Aristotle but is not nearly as intelligent). As an historical reference, it obviates the need for the further evolution of intelligence; a point also made by Robert Jastrow in *The Enchanted Loom.*

The influence of memory on religious belief is profound. Since memory is perception at a distance it can consolidate beliefs in reference to imaginary persons and actions. Because actual experience becomes irrelevant one can assign memory of human traits to imaginary figures, and create a whole new, extra-human, version of morality that allows for perfection in a way that concrete experience disallows. That can enhance moral learning.

Humans can be altruistic. They often come to the rescue, pitch in, save the lives of others and heal the sick. But they also murder, plunder, cheat, and steal. The storyteller/prophet might seek a righteous figure to impart a moral lesson in addressing those polarities but for that figure to obtain overriding authority he or she has above human status. That presents a quandary.

The solution? Simply take the benevolent aspects of human behavior and attribute them to a flawless figure, perhaps an angel, seraphim, or maybe part of a trinity in a heavenly firmament. The capacity of the imagination to mix and match enables us to create and acknowledge a variety of mystical entities.

Because memory allows for the attribution of flawed human behavior to flawless entities also makes it possible to create angels and demons as frames of reference by which to impart lessons on morality. However, that is only accomplished by taking the good and bad aspects of human behavior and putting judgment in the hands of a spirit who is perfect.

Is imagination-derived spirituality a superficial skill, a flight of fancy having nothing to do with the real world? Hardly. In fact, it is incredibly beneficial. The neural circuits enabling us to think, feel and act beyond the present through memory are responsible for advancements in art, music, education, philosophy, medicine, science, and many other areas. For example, many of the modern scientific concepts on space travel first originated from the mind of fiction writers such as Isaac Asimov, H.G. Wells, and even cartoonists like Philip Nolan and Dick Caulkins, who created the Buck Rogers comic strip.

Memory is a superb teacher of moral principles, which is one of its prime functions in the Bible, particularly through the intervention of angels and seraphim. Another mnemonic aspect of spirituality is seen in the capacity to imitate. It is well-known that imitation is a core part of human experience. In fact, it extends beyond that to primate behavior in general. The reason why is not certain but might have to do with the brain's computational functions.

When we pick up on a sensation, be it visual, auditory, or tactile, the source of the sensation is the result of an energy impulse. If it is visual - say a painting, the shapes and colors take the form of frequencies of light waves and angular shapes that impinge on the retina and optic nerve. In order to absorb that information requires a mirroring of those impulses via a neural signal correlate of the energy source. After that, the optic nerve sends signals to the occipital cortex, which can engage in any number of interpretations; for example, recognizing the painting as a distinct set of objects or landscapes, or rendering judgment on whether the painting was good or bad. Yet the imitation of energy impulses seems a primal aspect of sense perception and is so embedded in brain function that it typically leads to feelings of satisfaction. It is a recognition/closure response very much like Freud's "savings" component of humor.

That's why we laugh at impressionists, why we mimic the outfits of the Beatles, the gloves and moon walk of Michael Jackson, why kids buy Tom Brady's football jerseys and copy the hair styles and sayings of famous people.

That imitative trend goes beyond paying homage to notables, though the tendency to imitate them is stronger. It extends to all aspects of life, for instance, the sounds of animals in the use of onomatopoeia and the identification process in child development. However, imitation requires a capacity to internalize, and essentially incorporate into the self the entity one is mimicking. The extension of self to others spills over to the spiritual world. Thus, any spiritual entity conjured up in the human mind will always be a facsimile of something or someone else with which we are familiar.

As a result, we can and have turned people into spirits and vice versa. This is exemplified by the winged angels in Exodus 25:20, the spirit with a lion's head in Ezekiel 10:15 and other traits attributed to spirits that are similar to those of people. People are often viewed as soldiers - so are angels. People can announce forthcoming good and bad events - so can angels. People can applaud or chastise the actions of others - so can angels.

In that sense, the basic structure and function of the brain makes much of human experience a mirror reflection of what's around us. Our beliefs and personalities are shaped by things we observe and choose to copy. Spirits fit into that scenario because they are, in some sense, a reflection of the highest ranked humans: the ones we imitate and look up to.

Not surprisingly, the Bible is a vast source of references to the nature and importance of spirits. There are 103 references to angels in the Bible. Their roles were varied. Job 16:38.7 refers to them as Sons of God. In Psalms 89,6 and Samuel 1:11, 17:45 they are deemed the armies of God sent out to conduct spiritual warfare.

Another oft-mentioned role is that of messenger and middle entity between God and man, rendering service to those who will inherit the earth. It is as though while creating man in his image, God realized a certain amount of tutoring would be required for his most favored student to reach his potential. Therefore, in order to observe and mentor human beings, angels had to be higher up in rank, which meant being wiser and less bound by physical limitations.

Spiritual entities exist in virtually all faiths and are described in similar ways. The angels of Islam are said to be created from light, rather than clay, as with humans. That makes them formless and devoid of physical limitations. Their roles are similar to those in the Judeo-Christian faith. Indeed, the names of Islamic angels are derivative of those in the Old and New Testaments. For example, the angel Jib-Reel was derived from Gabriel. The angel Israfael is a derivative of Raphael. The angel Mikhail is derived from Michael the Archangel.

The Hindu faith also includes angelic entities. Devas and Atman are considered guardians and protectors. While their status as superior to mankind is similar to what it is in Judeo-Christianity, angels in the Hindu faith are accessible through meditation, rather than appearing strictly at the behest of God. In the Bhagavad Gita the communication between mortals and angels is ultimately for purposes of gaining a sense of unity with the universe, which has a pantheistic tinge.

An interesting aspect of this is how it dovetails with quantum physics, specifically the anthropic principle which assumes man is so completely tied into the universe that he cannot truly be an objective observer of outside events, any more than a single thread can be separate from a garment.

An interesting aspect of this is that scientists are also bound by cognitive templates. Regardless of how elegant their math models, experimental designs, or research findings they can never really answer the question of... *why*. They can only address... *what*. Indeed, it often seems that the more science progresses, the more it seems to head in a pantheistic direction. That ostensible lack of true objectivity makes the distinction and relative legitimacy of science and religion very tentative, which brings us to the issue of causation.

CHAPTER 7:
ETIOLOGY

A CENTRAL QUESTION REGARDING HUMAN NATURE is how a mind drawn to religion can also be attracted to a scientific methodology. One possible answer lies in the fact that the human brain has billions of inter-connective possibilities. The need to regulate arousal levels in such a large, complex brain makes closure quintessentially important. Closure can occur in many ways, but it typically involves a syllogistic framework. The syllogism is really a logic exercise. It involves an initial premise, several interactive variables and a conclusion based on how they affect one another. As discussed previously, a classic example is… If A is larger than B and B is larger than C, then A must be larger than C.

That statement is irrefutable. In fact, Greek thinkers used that method to prove various theorems. This was in part because the ancient Greeks did not have a precise math system. Like the Romans, their number and letter systems were intertwined. Syllogisms could be very accurate, as for example, Euclid's geometric theorems, which he never attempted to prove mathematically. Others, such as Aristotle's belief that the universe consisted of the four elements of fire, water, earth and wind were not accurate.

The key to a syllogism's accuracy lies in its initial premise. For example, since the statement... A is larger than B is specific, the rest would logically follow from that. However, if the initial premise was that A is sometimes larger than B, attaining closure would be more difficult.

As far as the mind is concerned, attempts to deal with initially confounding experiences should result in conclusions, even in creative circumstances. Forms of literature, painting, sculpture, humor, and performing arts can feature new twists on old themes, plots, characters, angles, colors, and punch lines, but creativity is not the same as chaos. To be considered creative, the format must lead to something definitive. In other words, as far as the human brain is concerned art, in any form, consists essentially of an extraction of closure from novelty - or uncertainty. This format is referenced in the ideas of many iconic intellectual figures.

In his discussion on humor, Freud referred to the uncertainty-closure sequence as "savings." By this he meant that humor entails a sudden recognition experience derived from an uncertain theme, for instance, an unexpected punch line.

The actress Mae West provided one famous example. Being single, she was once asked if she was opposed to marriage. She stated: "I have no problem with marriage. It's a wonderful institution. I'm just not ready for an institution."

The afore mentioned D.E. Berlyne referred to this process as the resolution of conceptual conflict. Some, such as Kellerman and Reynolds, discussed this from a social-communicative perspective and labeled the process tension reduction.

As with all mental experiences, the desired result is closure, which suggests any type of human cognitive process could be considered a form of art - an argument made somewhat flippantly by writer Jack Kerouac in an interview on the Ben Hecht show in 1958 when he stated that he wished to live in an America in which every man could be considered an artist.

Because closure is the necessary cognitive endpoint of human experience any number of beliefs and systems of thought can be adopted if conceptual finality is included in the process.

Science developed due to contributions from thinkers both modern and ancient, including assumptions by pre-human species. For example, the first species to make fire, Homo erectus, was able to manipulate nature by converting the threat of fire into a useful tool through closure. Homo habilis, ostensibly the first tool maker, also operated according to a closure-based template. He recognized that cutting a flint stone at a certain angle would sharpen it and facilitate cutting of meat and hides for food and clothing. Each of these primitive creations was based on a capacity to make empirical observations, a degree of calculation, an estimation of results and attainment of closure. This likely occurred first within the imagination, as the toolmaker envisioned what the outcome would be, then in actual experience once the tool began to take shape.

Much later, modern humans, including the ancient Sumerians, were able to make rough estimates about the movement of celestial bodies, and then attain closure based on further observations.

In subsequent times, philosophers now considered forerunners of science, like Aristotle, Galen, Archimedes, and Ptolemy came up with sophisticated theories that were internally consistent, even if not completely accurate. What sustained those theories was the "good fit" they provided between first guess and final draft. To that point, knowledge was partly objective based on empirical observations and partly subjective, based on closure. Eventually that changed.

The true originator of the modern scientific method was Rene Descartes.' He provided science its most precious variable, humility, through his principle of universal doubt. This is now referred to as the null hypothesis. It created a rigorous standard adopted by scientists who would start out by assuming their hypothesis was wrong.

Consequently, rather than hoping the theory was correct prior to experimentation, they would describe success as "disconfirming the null hypothesis." In other words, the results overturned the initial premise that the theory was wrong. This marked the true beginning of what came to be known alternatively as the empirical philosophy and the philosophy of science.

As the empirical movement gained momentum, the importance of philosophy and theology diminished in some quarters. That trend irritated philosopher Immanuel Kant so much that he felt offered a strong rebuttal in the statement: "Perceptions without conceptions are blind." By this he meant one could observe a phenomenon until the end of time but given the proclivities of the human mind, interpretations of those phenomena would have to be framed by the templates of mind. It was a notion much later championed by Steven Hawking.

Hawking, of course, believed wholeheartedly in the value of mathematical proofs. Still, both felt the mental templates of mind had to be factored into any conclusions about the natural world. In fact, there is only one distinction to be made between philosophy and science. The latter relies on math, while philosophy relies on logic, more specifically the syllogism, to prove the accuracy of any given hypothesis.

In ancient times, during in the Bronze and Iron Ages and leading up to the 9th century, math was not precise enough to make accurate predictions about the actions and nature of objects and forces. Yet there was still the need for closure. The brains of Abraham, Moses and Jesus were every bit as large and intricate as those of Isaac Newton or Albert Einstein. Just because they could not use a statistical analysis to measure physical relationships didn't mean they were unable to draw conclusions about the world around them. Indeed, given the structure and functions of their closure-seeking brains, they would have had to.

One of the ways they did this was by resorting to a pre-scientific method often referred to by Biblical scholars as etiology. The word means 'origin' - or cause, but for pre-scientific peoples, the cause-effect sequence was reversed.

In modern, post-Cartesian times we believe stimulus A causes outcome B. For the most part A is thought to precede B. Not all modern scientific methodology adheres to this time sequence. For example, behavioral psychologist B.F. Skinner proposed that behavior is a function of its consequences. By this, he meant a reinforcer (or reward) provided after a behavior, would cause the behavior to become more frequent, i.e.to be learned. Another reversed causative sequence is found in Darwin's theory of natural selection, which proposes that genetic traits are either selected or discarded by nature after the mutation occurred. Regardless of one's point of view, cause is usually presumed toprecede effect.

Conversely, the etiological perspectives of ancient peoples were based largely on prophecy. At times, societies did look for omens in anticipation of a good crop yield or military victory. They undoubtedly carried out their everyday lives according to a traditional cause-effect mindset. For example, Jesus's father Joseph, a carpenter, surely knew that in building a table he had to piece together basic step by step wood cutting and fastening sequences before completing the job.

Yet, when it came to religion and other spiritual matters, causation was often assumed to be predetermined. After suffering a loss in battle, tribes tended to explain the outcome retroactively, as though the setback was already decided by God, irrespective of tactics, motivation, or other causes. One could have a superior army, more sophisticated weapons and an efficient strategy but still be destined to lose as a result of faithlessness.

It was an interesting model of attribution with both cost and benefit. While it ceded control to God, it provided irrepressible psychological resilience. For example, losing a battle would not necessarily lead to a sense of incompetence or futility. Instead, it could provide a guideline to future success, usually attainable by paying proper homage to God. With that as reference point, a David could slay a Goliath using just a slingshot

The logic, though a bit awkward, was fool proof. In order to win the next battle or reap a better crop in the next harvest, one only had to commit more fervently to faith. Whether the opponent had better generals or were better prepared was not the point. Loss was related to sin, victory, to the moral purification of the individual and the tribe.

The impetus that provided would have been phenomenal. With each setback, the tribe would seek to become truer to their beliefs. In that context, it is not surprising that religion was so paramount in the lives of these people, and why monotheism was so functional. Having one God who ostensibly favored a specific group of people, would narrow the focus on laws and moral principles, make adherence to religious beliefs more facile, and ensure greater and success in war, farming, fertility, and every other aspect of life.

While unusual – particularly in comparison with the modern world, that etiological mindset provided a tremendous boost to culture, morality, production, and even mental health. Attributing all important events to the proactive decisions of a higher power took the weight off the people's shoulders. Despite lacking an empirical approach to religious matters, mathematics, social policy and medicine, the ancients survived and, in many instances,thrived.

Their capacity to adapt, despite having such a non-empirical mindset attests to their adaptiveness and their ability to weave cognitive skills into a functional mosaic. As long as they could obtain closure, even in a retroactive, attributional framework, life would go on.

That makes the evolution of human cognition a fascinating topic, because in many ways, the structure and functions of the human brain can insulate humanity from nature's random decisions. As alluded to previously, in evolution, neurons in the frontal and prefrontal cortex expanded into a functional vacuum less concerned with sensory or motor faculties than with plans, predictions and the imagination - in other words, with events that had not yet occurred.

The sheer mystery and plasticity of the frontal cortex enabled the afore-mentioned Phineas Gage to function normally despite severe damage to his frontal cortex resulting from a railroad accident, andprompted writer

Robert Jastrow to surmise that humans will not (and need not) experience further evolutionary changes because cognitive flexibility enables us to beat nature at her own game. The idea was that while, through natural selection, nature tests other organisms, it appears frontal expansion has enabled humans to steal the answers to the test beforehand.

Whether true or not, it does seem we have been able to overcome at least some of nature's environmental pressures in ways no other animal can. Is this something that can be understood in a religious context?

Is it possible, despite the pantheist belief in oneness, that humans have been handed the reins of the natural world by evolution and by God. In that regard, some intriguing questions can be asked as to whether Homo sapiens' concept of God is really a glimpse in a mirror, and whether God is physiologically, neurologically, and genetically within us.

In his emphasis on objective science, Descartes also wrote something that was highly conjectural. He proposed that the soul of man was housed in the pineal gland, a small structure located in the midbrain. While Descartes' physio-spiritual idea was wrong, one could ask if there is a God-circuit located in the frontal cortex that makes religious worship not only possible but unavoidable.

CHAPTER 8:
KINGS

THE FIRST REGAL AUTHORITY IN THE HEBREW NATION was Saul. The fact that he was anointed king was both unifying and disturbing to the tribes of Israel. As if anticipating the need for checks and balances on power seen in modern democracies, some Israelites felt it necessary to limit his power, especially as pertained to whether his power would lead him astray of religious tenets. Because of that, it was left to the prophet Samuel to provide restraints on Saul's power. He did so in a speech to the Israelites in which he presented a list of warnings.

Samuel was not speaking for all the people. Many believed establishing a monarchy with God's sanction would be beneficial. Their reasoning was historically justified. In the aftermath of endless battles for territory and resources, small tribes were being obliterated militarily and culturally. It occurred to some that the new cultural survival strategy lay in numbers. The nation with the most soldiers and weapons would dominate, while the smaller entities could end up relinquishing resources, native languages, and customs.

Samuel was prescient enough to know that while anointing a king would allow the monarch to build a vast military and sustain the nation, he also knew the power accrued by a monarch could lead to tyranny and religious and cultural discord. His cautionary speech is referenced in 1 Samuel 8:10-18

And Samuel told all, the words of the LORD unto the people that askedof him a king.

And he said, This will be the manner of the king that shall reign over you. He will take your sons and appoint them for himself, for his chariots and to be his horsemen, and some shall run before the chariots.

And he will appoint him captains over thousands and captains over them all and will set them to reap his harvest and to make his instruments of war, and instruments of his chariots.

And he will take your daughters to be confectioneries and to be cooks and to be bakers.

And he will take your fields, and your vineyards and your olive yards, even the best of them and give them to his servants.

And he will take the tenth of your seed and of your vineyards and give to his officers and to his servants.

And he will take your menservants and your maidservants and your goodliest of men, and your asses, and put them to work.

He will take a tenth of your sheep and ye shall be his servants.

This series of warnings led to regulations, and subsequently to ongoing conflict between Saul and Samuel throughout the former's reign.

Here, Samuel was simply stating the obvious, and anticipating a point made by Abraham Lincoln and Thomas Jefferson: that the real problem was not the presence of a king, but traits within human nature leading to the quest for power that could lead to corruption and tyranny.

He might have been referring (whether consciously or not) to the purity of the nomadic tribe in his admonishments. Once upon a time, there was social, genetic, religious, and economic solidarity. Every 't' was crossed in an age-old social system, in which work, and love had to be shared and familiarity among members led to a fluid group dynamic. He probably realized the times had changed and that nation building had become necessary due to a thing called war. However, as a spiritual man, he was inclined to look beyond transformations in culture and politics. Samuel's misgivings would crop up throughout the history of our species, including up to modern times, He protested the confiscation of property and disassembling of the family. So did the Sicarii in response to Roman acculturation, the English prior to the Battle of Hastings, the American colonists in response to being taxed without representation, and the French at the Bastille. It might be accurate to say that most revolutions resulted from misgivings mentioned in Samuel's speech.

The presence of a king has presented a dilemma for almost every sovereign entity that ever existed. The reasons for both needing and resisting regal authority are complex and likely date back to times when human tribal conglomerates invaded territory with mercenaries and extra-familial combatants.

One reason has to do with group mathematics. Two factors likely created tension as human societies expanded. For many thousands of years there was congruence between the social patterns of our species and our genetic (or at least natural) social dispositions.

All creatures on earth tend to prioritize the welfare of their genetic relatives – and so did our ancestors. In that regard, Richard Dawkins, author of *The Selfish Gene* introduced (and defended) the idea that much of our behavior is directed by the actions of genes, more specifically the DNA molecule.

Yet a biochemical does not have a brain and one could assume it lacks the capacity to make decisions. Genes obviously dictate where, and how proteins are aligned to produce specialized tissues and organs but extending that to the execution of behavior patterns seems to stretch the imagination.

On the other hand, genes do some things that can be described as decision making. They seem to "know" how to assemble protein chains and how to correct errors in the transmission of RNA to DNA. They have a quasi-cognitive ability enabling them to line up the four base chemicals in the right order so that guanine, cytosine, thymine, and adenine end up connected to the right partner. When one considers the complexity of this process and the few errors that occur within each generation of life forms the mechanism seems more than cognitive. It seems miraculous.

Whether or not genes dictate behavior patterns, it appears their impact on an organism's overall functioning is profound. Indeed, the lower one goes in the phylum - assuming one can arbitrarily determine lower and higher status among organisms (bearing in mind the prime measuring stick for success is adaptability, and that ants, beetles, spiders, and crocodiles have existed much longer than humans) the more genetically programmed their behavior. Fixed behaviors like the spider's weaving of webs, the leaf cutter ant's agricultural capacity to harvest leaves and the bumblebees' distribution of pollen are fixed, instinctive patterns.

One reason they are fixed is because their brains are small. As brains expand, more circuits provide more volume, along with a greater range of associations. In the case of humans, that produced a capacity to partially override primal instincts with learned responses and abstractassociations.

It is conceivable the human tribal, egalitarian social model would have been multiply reinforced by both instinct and learning. First, it could have been driven by a genetically engineered tendency to continue the line. Secondly, it could have been reinforced by emotional comfort due to social closeness, behavioral predictability and motivational benefits resulting from interdependence among members in a familial/nomadic group.

In the small tribe, group cohesion would have been maximized. While primate groups tend to be hierarchical, the size of the group, and tendency to preserve the genetic line would have led to limited numbers. That would have led to a balanced division of labor, and to a large extent minimized the need for a hierarchical (class) social system.

The modern human race has probably existed for 250,000 years, as conveyed in Allen Wilson's Mitochondrial Eve theory. That means for over ninety percent of our existence we functioned and adapted according to group numbers no greater than two hundred members. Once that pattern changed, there was bound to be stress, desperation, and a search for solutions to the problem of discomforting social and genetic diversity.

Over time, as nations expanded, kings began to seek more power than governance required. Part of the reason was societal tolerance. Part of it was evolutionary. Human societies can claim to be egalitarian, and certainly aspire to that in areas like law and education, where equal rights are codified. But in the primate world, of which we are part, social rank is natural. Alpha males and females are an uncomfortable necessity.

This trend is prevalent in every human society. Authority figures and celebrities receive special privileges. Once in power, they can intimidate and bully "lesser" members. In many instances, the law treats them differently, members of the opposite sex treat them more favorably, and they are excused, even idolized in the aftermath of their transgressions. Yet, while those are unfavorable aspects of fame and power, icons are also necessary signposts within a culture. Without idols, we could have no heroes, no entertainment, art, or politics.

In human experience, alpha males and females are more important than in other primate groups, because they have symbolic value. Also, we humans have such vast memory capacity that we can keep them around for historical, educational, and inspirational purposes for as long as we want – people are still writing about Lincoln.

On the other hand, we are ambivalent about alpha males and females. Why? Conceivably it is due to our social intelligence.

Humans can perceive details like no other species. As a result, it is easy to find flaws in our heroes. The flip side of honor is blame. Heroes are given exalted status due to achievements that are considered above the norm. But since they can excel, they are expected to come through every time. If they fail morally, physically, or intellectually, we are quick to dismiss them, and at times, persecute them.

In choosing Saul as their king, the Israelites might have considered the benefits of power more than the drawbacks. That is understandable. Their situation was desperate. Therefore, with Samuel's blessing, and after a few losses in battle to the Philistines they eventually decided to go headlong into nationhood.

The ambivalence among humans regarding kingship seems to reflect the conflict within human nature on whether to accept or resist the idea of a social hierarchy. Whether a king, an administrator, mafia capo or football coach it seems the polarity is palpable. From worship to disdain, from honor to disgrace, from the top to the bottom and in some instances (for example in 18th century France and 20th century Russia) from a life of luxury to a violent death.

Despite gaps in recorded history, and before writing was systematized, it is generally believed the era of kings began with Sargon of Akkad. His reign began around 2334 B.C. Like most kings, he built his reputation on military victories. In his case, it was the conquest of Erech. Interestingly, Sargon was a Semite, and like another well-known Semite named Moses, he was reputedly abandoned as an infant and placed in a basket which was discovered floating in a river.

However, while Moses was discovered by an Egyptian woman Sargon was introduced into history by Ishtar, the Sumerian goddess of love. After a series of conquests Alexandrian in scope, he pronounced himself king. He did so in a way that was emulated throughout the course of history. He insisted his kingship was granted through his communion with the Gods. It was a brilliant ploy that doubly reinforced his power.

As discussed above, there has always been a tendency for people to grow tired of monarchs. After all, they are just men and women, and in the final analysis, one impetus for creating various gods over time was the need for an authority figure who was not flawed, wasn't flesh and bones, and had no "stake" in the material world. That kind of authority would not abuse his or her power by overtaxing, stealing, plundering, and enslaving subjects.

By combining the two leadership models, Sargon was able to rule for fifty-six years and leave a dynastic legacy that lasted over a century. During his rule, the Sumerians developed the cuneiform writing system.

As with all historical accounts, the record keeping back then was less than completely objective. However, there is no indication that Sargon was cruel or abusive, at least to his own people, and while conquest always entails bloodshed, it seems once his domain in various locales was established his governing style leaned more toward trade and cultural development than brutality. Despite that, his man-God fusion was rife with potential tragedy and would become one of the most dangerous ideas purveyed in human history.

In a sense, it diluted the purity that came along with separation of God and man. With his gesture, there was no longer a holy overseer. While in later times Samuel adopted a compromising role in dealing with Saul, that was an anomaly. For the most part, monarchs continued to claim close ties with the gods, which gave them license to run roughshod over the people.

Not that there weren't benevolent kings along the way. Perhaps, sensing the potential for endless power grabs, Hammurabi of Babylon created a legal system he hoped would provide honest detachment. He developed a codification of dos and don'ts, and this was the first document with the potential to stem the tide of oppression. Fortunately, his legacy would become even more influential than the god-king model.

The code of Hammurabi is an interesting document on many levels. It appears to have been the first formal set of laws, although the Sumerian chieftain Ur Nam mu created a precursor in 2100 B.C. In reading Hammurabi's text, one gets a sense of the desperation involved in organizing and sustaining a heavily populated nation state. His reign was much later than that of Sargon - which was roughly between 1792-1750 B.C., but the content of his writing suggests there was considerable duress involved in transitioning from nomadic to urban society. His code consisted of 282 laws which he divided into three main sections that were similar to laws in modern western societies. The three components were procedural law, property law and the law of persons.

Although dogmatic, his attempt at jurisprudence was remarkably consistent. One of the most striking aspects was its devotion to proportional thinking. Numerous laws pertained to relationships (particularly marriage), commercial transactions and the bearing of false witness. There were four levels of punishment and three levels of mitigation. One punishment was by drowning, whereby people would be tested by being thrown into water to see if they could survive. If so, they were presumed innocent. Another form of punishment consisted of being burned to death. There were also corporal punishments that fell short of death. For example, if a man put another's eye out, his eye would be removed. If a man broke another man's arm his arm would be broken. If a son was born to his father and a prostitute and denied his parentage his tongue would be cut off.

Such punitive measures were meted out for acts such as falsely accusing another person, striking one's father, stealing cattle and kidnapping, i.e." banning" someone unjustly. Another penalty consisted of paying a fine for deceptive financial transactions. While there was an attempt at fairness by the king, Babylon featured a class system that treated slaves and freemen differently. A similar disparity existed between men and women.

In modern times, some feel it necessary to condemn the acts of historical figures that are at odds with modern morality. However, all people and generations act, think and moralize according to the circumstances in which they live. In that context, the most reasonably moral individual of any era is not one whose reasons coincide directly with modernity but one with moral and/or political innovations that surpass the trends of the times.

In a sense, Hammurabi reached that threshold. Like Madison, Jefferson, and Moses, he humbled himself before God. His legal code contained a framework similar to subsequent bodies of law, including the Ten Commandments, the Roman Decemvirs, the Magna Carta, and the American Constitution. While giving himself authority to write the laws, he attributed his wisdom to the gods Anu and Bel, consistently referring to them as ultimate decision makers. In fact, a preamble to the text describes Hammurabi as "the exalted prince who feared God."

Conceding primacy to the gods enhanced Hammurabi's authority. He wasn't claiming to be a god, nor asserting, as would future leaders, that he and God were so intertwined that defiance to one necessarily meant defiance to both. His concession was an act of genius, particularly given the tendency of the Egyptian pharaohs and later, emperors of Rome, to equate themselves with God. In effect, Hammurabi was working on a system he hoped would be person proof - a transcendent legal code that precluded tyranny.

He did not quite get there. Some of his laws were absurd; for example, a physician who erred during treatment might have his hands severed. Also, his was a class system, whereby acts against slaves and the poor were not as severely punished as they were against freemen, administrators, and religious leaders. There was a value scale among people in Babylon and Hammurabi did little to alleviate that problem. However, his sense of fairness did hint at the possibility of change. Like the framers of the U.S. Constitution, he created laws allowing for upward mobility; for example, by allowing women slaves to marry freemen and administrators and to inherit all the rights and privileges inherent in a typical marriage. That included enabling a wife from the slave population to inherit her husband's property upon his death and to be treated as a free person herself.

Hammurabi's relative benevolence toward women did not extend to prostitutes, although his laws stipulated that anyone born to freemen and prostitute who chose to deny his parentage could be held criminally liable. Thus, while dogmatic and a bit heavy on the death penalty, Hammurabi's code provided a first step on the road to fairness in human society.

One aspect of his character seems clear. Hammurabi was a perfectionist who knew maintaining order in a large society was going to be difficult. To have peace and tranquility required more than a king. It also required the supervision of gods over kings, as well as a series of laws that could not be manipulated to suit the whims of tyrants.

Moses came several centuries after Hammurabi. He was born somewhere around 1323 B.C., died in 1272 B.C. Some have suggested Moses' words and actions were similar enough to the ideas of previous figures to appear derivative. At face value, the argument has merit. There are similarities between his first appearance on the scene and that of Sargon. As discussed above, both were supposedly placed in a basket as infants and discovered floating in a river. Meanwhile, like Hammurabi, Moses deferred to God as originator and ultimate authority regarding laws in the Torah. However, Moses' theosophy signified a leap forward in socio-moral thinking.

Hammurabi's code was practical. While attributing its authority to the gods, his system was primarily secular. Moses's laws were far more spiritual. Also, unlike Hammurabi, Moses did not describe himself as "exalted" and deserving of any sort of primacy. In fact, he was reluctant to assume the role of lawgiver, expressing his ambivalence to his god in Exodus 4:10 by asserting he was not glib enough to assume that heavy responsibility.

The shift from a practical, secular legal-moral model to one more spiritual was only part of Moses' contribution to modern thought. His ultimate perspective differed from Hammurabi's in more fundamental ways, which illustrates how times changed over two centuries and was indicative of the evolving mores of Hebrews at the time.

Although issued with the implied sanction of the gods Hammurabi's laws were concerned with economics and civic issues like marriage, slavery, medicine, and obedience to authority. Moses was simply trying to summon God's will, as if from child to parent, so that He, the Lord, rather than an exalted king, could step in, establish peace, put an end to bondage and provide prosperity for the tribes of Israel. Hammurabi was conservative, determined to preserve order while Moses was a revolutionary opting out of the existing society and with God's help, intent on creating a new one.

For those reasons, the laws of Hammurabi and those imparted to Moses were bound to be different. After all, Moses had been reduced to the status of Egyptian slave in Exodus while Hammurabi was a king. Moses initially had a right to feel as exalted as Hammurabi despite being "slow of tongue." Also, The God variously called Yahweh and El had not made many appearances over time and those with whom he did make contact could easily be described as exalted.

Yet, instead of riding the wave of opportunity Moses chose unconditional faith and humility. This was extraordinary, considering that Moses had been a significant figure in Egypt before reconnecting with his people. Why the choice to descend in rank? The answer lay in a moral concept initiated by Moses that emphasized compassion.

People in the ancient world had the same brain structures as we do today. However, they seem to have been "wired" differently regarding social sensitivity. While killing was punished in all ancient societies, killing and acts of extreme brutality were common and much more accepted than is the case today. A person who stole could be put to death. A woman who committed adultery could be killed by stoning. When a free man died, his slaves were often sacrificed to accompany him to the afterlife. The fact is death was not considered so dreadful in ancient times.

There were probably several reasons for that. People back then had to deal with high death rates from disease, predation, and combat. After witnessing brutality after brutality, the prevalence and devastation caused by plagues, lack of sophisticated medical treatments and high infant mortality rates, people living in those times probably learned to adapt emotionally to the specter of death. Consequently, they were able to carry on more readily in the aftermath.

In an evolutionary context, another factor might have come into play - the feminine voice. Humans have a genetic similarity to other primates. We are not descended from apes but seem to have evolved from a species ancestral to chimps and humans.

Many paleoanthropologists believe the common ancestor was the species Ramipithecus, a flat-toothed creature whose concave pelvic structure might have created the anatomical prototype for bipedal ambulation. It is possible humans and chimps branched off from this forerunner about seven million years ago.

In that context, there is an interesting behavior pattern typical of primate females. They spent a fair amount of time intervening when males become violent. In some primate groups this intervention involves soliciting males for sexual activity in the hope that copulation will calm males and prevent harm. In other instances, high ranking females will try to separate combative males. It often works, as it does among humans. This tendency has interesting bio-social roots.

It is well-known that the relative status of males in any primate social group depends on the importance of aggression; not just in defending the group but also in controlling populations and for hunting and engaging in physical labor. As unpalatable as aggression might be to us, it was critical in sustaining primate and early human culture.

Aggression became less important as human society became more sedentary and technological, which raised the status of females. Factors such as control, domination and physical prowess were gradually obviated by the invention of machines. That led to women having a stronger voice, which meant their criticism of aggressive behavior began to carry greater weight.

In ancient times very few societies treated males and females equally. Moreover, a female's choice of mates was restricted by familial, rather than personal decisions. That minimized the impact of criticism by women, which freed up males to act more in accord with their competitive, dominance-oriented proclivities. All such factors had to be involved because there is no "aggression gene" forcing males to fight and kill. Indeed, some of the most humanistic acts of compassion in human history have been by males.

Still, left to their own socio-hormonal devices males would tend to act in ways that increase the level and frequency of aggression in any given society. That is why there might be a close correlation between the status of females and attitudes toward aggression in various cultures.

Beyond that, death was not necessarily viewed as final in ancient times. Almost all human groups - perhaps beginning with Neanderthal, believed in an after-life. As mentioned above, slaves were often sacrificed to accompany their master into paradise. In many instances, rather than resisting, the slave considered this an honor, especially since the afterlife was usually described as a place for the wealthy and powerful. An unwavering belief in ending up in heaven, Elysian fields, Nirvana or some other Utopian vista could have overcome the impulse to survive.

The rise of the feminine voice ameliorated that over time. Yet, if, as Freud, Jung and others suggested, there is a neuro-behavioral pathway leading from the womb to the mind to culture, in which women's preference for the creative - libidinal (as opposed to destructive - thanatotic) aspects of life, including the arts, social interest, nurturance and giving birth carried great influence, that influence had to lay in wait until mechanical power was invented. Prior to that, systems of servitude made death by sacrifice less odious to laborers and slaves. That would not change until someone came along to challenge those systems, to make life more important, and to make death less acceptable

In that context, Moses stands out as a singular figure. He was able to feel for the poor and enslaved, to internalize their suffering despite living in an era of male dominance, and he attained this sensibility well before encountering his God. Egyptians overworked, abused, and neglected their slaves. Moses' initial protest was against mistreatment of them, even though he enjoyed high status in Egypt. While his God orchestrated an escape from Egypt it seems Moses' purity of heart preceded his ascent on Mt. Sinai.

Indeed, in some ways the God he encountered seems to have been less empathic than Moses. The word of God in the Decalogue was absolute. No transgressions would be tolerated, even by Moses' own people, who were punished for disobeying the first, and most binding commandment: *Thou shalt have no other gods before me.*

When Moses brought the people out of Egypt, the unruly, impatient masses became frustrated. They contemplated abandoning Moses, questioned the power of his God, and threatened to return to Egypt. In response to this, God forced them into exile and sent vipers to kill someof them.

In that sense, there seems to have been a difference between the gentle advocacy of Moses toward the Hebrew slaves and the black and white actions of his God. In later epochs, Judeo-Christianity would move toward love as a staple of faith, but the initial template of the Hebrew faith seems to have been based primarily on obedience.

Did this signify that the god of Moses was cruel and unforgiving? If so, why did Christianity, as a Jewish derivative, come to emphasize love, forgiveness, and repentance rather than punishment? One could speculate about that in the context of human nature.

Moses' people were of mixed ties and beliefs and not yet united behind a single religion. Though many believed Abraham was founder of a monotheistic faith the latter's conception of God was general. Abraham was raised in a culture with strong elements of pantheism and paganism. His conversion was likely based on a sense that a religion consisting of gods scattered here and there, regulating various and separate functions was insufficient.

Moses's task was much more difficult. His people were eager but had to wait. They were disorganized but had to act as a cohesive group without really knowing much about the word or the identity of Moses' God.

This God had no name, indeed could have no name as indicated in the episode of the burning bush. The comment... *I am that I am* in Exodus 3:14 was an utter rejection of Moses' request to categorize God in any way.

Converting such neophytes required a firm approach. It was much like the training of a young soldier who comes into the military lacking in discipline and requires indoctrination. Turning Moses' followers into soldiers of faith would require rigorous training. Despite his compassion for the slaves, Moses understood this. The discipline required to build and sustain the Jewish faith was possibly one reason it took centuries before Jesus of Nazareth converted faith into a more compassionate model, in which prostitutes were forgiven rather than stoned, in which tax collectors were invited to dinner and a Roman centurion's request to cure his servant was granted.

Still, it seems Moses was not only a lawgiver but originator of an altruistic template, even if he knew harshness was needed in the beginning. His ultimate solution to this dilemma appears to have been a retreat to the familial bliss of the nomadic tribe, whereby social cohesion was really the most important command. He had to build a tribal mindset to fortify the faith under the guidance of a cohesive set of laws. This strategy is illustrated in Moses' final address to the people in Deuteronomy 17:14-15.

When you come into the land in which the Lord is giving you and inherit it and live in it, and you say, "Let us appoint over me a king like the surrounding nations, then you will appoint over yourself a king whom the Lord shall choose. From among your brothers are you to appoint over yourself a king. You may not appoint over yourself a foreigner who is not your brother.

This passage is so consonant with Samuel's tentative sanction of Saul's kingship that it seems to have reflected a deep feeling (and warning) suggesting that while nationhood was necessary, it was not preferrable. While historians point to the benefits that arose in the first agricultural societies, many people at the time felt it jeopardized their moral integrity and family structure.

It reflected a theme that persists to this day. For example, many libertarians and conservatives believe

that increasing the scope and power of government will lead to disintegration of the family, and that a resilient society is one that resonates morally, socially, and emotionally from the core family unit.

One of the more interesting aspects of Moses' behavior has to do with his para-deistic beliefs. He was, after all, more than a lawgiver. Accounts in the Old Testament (Exodus 3:1-10) suggest he was 76 years old when he received the Ten Commandments. As with Jesus, it is important to consider his early life as a determinant of his subsequent actions. His life was more definitively documented than that of Jesus, but the question remains; what was the learning process that brought him to God's will?

Movies about Moses portray him as an innately good man who became appalled at the treatment of Hebrew slaves, even before he knew he was Hebrew by birth. It was supposedly his defense of slaves that got him into trouble. His acts of compassion were viewed by some as acts of defiance against the Pharaoh, Setti I. Yet the Old Testament suggests Hebrews had been enslaved in Egypt for centuries and were treated horrendously for the duration. Why then did Moses choose to veer off the beaten path?

It seems, in the quest to record history and understand the nature of God we often neglect to analyze the personalities and motivations of prophets, law givers and other transformative figures. Yet, Moses was a man. Thus, it seems fair to ask what he was like, aside from his devotion to God. Perhaps he was like every other human being who ever lived, in harking back in mind and soul to a homo-indigenous time, place and set of circumstances. His ancestors roamed the planet Earth for a quarter of a million years. Their social groups were small but highly functional. They did not have trucks, planes, or chariots, and they could not accumulate wealth. Hording was not sinful - it was inconceivable. Greed was impossible, and the need for contributions from every single member, and for the group to be vigilant about caring for the young created an intensive level of social camaraderie. It was a bit like the economic principle of inflation...the more money floating around, the less its value.

The same principle could apply to people. No one is expendable in a nomadic tribe whereas many people become expendable in a heavily populated environment typified by redundant skills and competitive desires. Indeed, one could argue that democracy was not invented by Greeks or Invented at all but is an ingrained bio-social trait emanating from a brain programmed to respond to population dynamics.

In that context, I believe Moses' task involved more than a search for God's word. His quest was perhaps no different from that of any natural man. He was trying to replicate pro-social tribal dynamics in the framework of a densely populated society and to turn divisiveness into cohesiveness. His response to the abusive treatment of Hebrew slaves might have manifest a primal social instinct housed within the confines of mind.

Is there a tribal instinct within the human genome? It has never been proven scientifically, but evidence can be assessed in various ways. One of which is through mechanisms of behavioral homeostasis, or what could be called internal regulation.

As an example; human metabolism functions normally within a certain temperature range. When the temperature falls below a certain level the body will shiver as a heat retaining adjustment to the cold. If the temperature is too hot, the body will perspire and summon other cooling responses. The question is whether these physiological mechanisms extend to broader social behavior patterns.

There is a proven positive correlation between crime rates and population density. The denser the population the greater the tension level, in part because competition is more fervent for jobs, mates and general opportunities. In a socio-biological context, all animals react emotionally to overly dense populations. They instinctively recognize that access to resources needed to survive and propagate depends on how much territory is available to them.

For various reasons - all ultimately tied to survival - individual and group tension increases in proportion to population density. This also applies to the degree of genetic diversity within the population, possibly because diversity threatens the genetic purity that is presumably "selected" by specific gene pools. That is not to say humans are inherently racist or xenophobic. It does mean genes engage in a certain amount of self-serving behavioral inducements that we must, through our cognitive capacities and pliable minds work to override, in order to overcome bias.

There are bio-social reasons for preferring a smaller population that have nothing to do with race or ethnicity. Disease is transmitted more readily in heavily populated settings. Most creatures can assess the relation between population and available resources. This perceptual understanding is often referred to as territoriality, and it is a reaction to the possibility that an influx of invaders could result in depletion of resources. Most animals defend their territory as a survival mechanism. It is apsycho-biological process that works its influence through emotions and behavior patterns. In other words, in the overall scheme of things, dense and diverse populations can signal danger as well as foment anxiety, aggression and avoidance patterns - including for our species.

If Moses was acting (even unwittingly) according to a belief in a tribal culture, catering to dominance by pharaoh - even one who had treated him like a son, could only have been done out of desperation. The emerging trend in the Middle East had been urbanization, largely due to the need for armies and the demand for resources. Trying to maintain a cohesive social system was fine. However, the entire area was fraught with violence, conquest and ever-increasing sophistication of weaponry and military tactics. Small tribes had no protection against that. In most instances, tribals would meander into existing sovereignties to find work and set up a domicile. However, in so doing they would come under the control of monarchs or civic administrators.

Those in power wanted to embellish their status so they built great monuments to themselves. Labor was required, and the tribals were conscripted to do grueling work needed for construction projects that took decades to complete. That meant people were forced into hard labor, with little compensation. Furthermore, much of their earnings (if they were considered "freemen") was turned over to tax collectors.

Back then, sovereignties operated much like modern organized crime families. You had to pay and sacrifice mightily to be protected from invaders. Options were limited, While Pharaoh might enslave you a successful invasion by the Hyksos, Babylonians or Canaanites might result in the slaughter of your entire family or, in more "humane" circumstances, confiscation of your property, your wife and your children. In essence, there was a vexing political conundrum back then. One could remain tribal and at one with the natural world and risk devastation and death, or join a formidable nation and survive, while enduring humiliation, suffering and poverty.

Moses was a point man and intermediary in all this. His task was more than religious. He was not just a lawgiver, but a man of his time who was faced with the same realities that his eventual successor, Jesus of Nazareth would encounter centuries later. He must have known that even with complete trust in his God the search for a land of milk and honey would require more than commandments. One had to first enter that land, fend off intruders and overcome or bargain with prior inhabitants. One would also have to maintain order among people who were united only tentatively, and if faced with group duress, might develop internecine rivalries despite the word of God. Therefore, notwithstanding misgivings about the new mega-social experiment Moses and his flock did embark on the path to a modified version of tribalism. The task was to find a way to combine the altruism of the tribe with the strength of a nation.

It might have begun with the prophet Samuel's reluctant agreement to anoint Saul as their first king. Saul was not a terribly effective monarch. He was trying to juggle the role of king with that of religious spokesman. It came across as weakness and arrogance. The people wanted him out because he was not formidable enough. While his military campaigns were occasionally successful, he seemed incapable of fitting the mold of king, which among the fledgling, nation-building Hebrews was vaguely defined to begin with.

The prophet Samuel anointed him reluctantly and only at the request of God and the people, as written in 1 Samuel 8:6.

Saul must have felt great trepidation about how to act. He had been around. He knew how kings in various sovereign ties exercised their power. Saul could not be that kind of king - Samuel would not allow it. Saul was like an interim football coach who was not allowed to choose draft picks or make out the starting lineup. He was a figurehead with limited power who was fired because he was not powerful enough.

Saul knew submission was the psychological default position of his people. His God was unforgiving, and as a monotheistic figure, the only heavenly consultant to whom one could turn. One wonders if Saul blamed Samuel and even God for copping out. If they wanted a strong king, why the tone of reluctance? If God was the one and only, why pass the buck to a king by making him responsible for caring for the poor, defeating all enemies and somehow acting strong and weak?

At first, he did not perform badly. Saul's legions won battles, including against the rival Amalekites. However, despite God's command to wipe them out Saul let up on them. In his humble heart he believed they had had enough. He didn't understand their fervent commitment to win and fanatic devotion to war. They acted Spartan and Saul acted Athenian.

The Amalekites eventually returned and won. God got upset, the people got upset, Saul was deposed and although his only son, Ishmael attempted to sustain a family dynasty, Saul's rule came to an end. Some accounts have him killed in battle. Others suggest he committed suicide. In any case, the first Jewish experiment in secular rule and nationhood had failed.

Yet, human nature is what it is. Regardless of how spiritual the people, losing a war is unacceptable. Once again, one of the last tribal entities found itself straddling the fence. On one hand, they wanted a religious purist to lead them. On the other hand, the next leader could not be so pure as to be weak. He had to be aggressive, and a winner who would not co-opt the power of the Lord.

There was great trepidation after Saul's death. The question revolved around whether there was anyone on this earth who could be both strong and submissive to God, both humble enough to win the hearts and minds of his people, yet confident enough to run the show. It represented a Jewish cognitive template that would last through the ages: the image of an ideal leader who was both weak and strong, both a suffering servant, and one so transformative as to reshape the entire family of man.

Their search for a politico-religious leader led to the reign of King David. He was an attractive combination of shepherd boy and gifted musician who was sensitive, in some ways meek, lacking the temperament of a tyrant, yet one who slew Goliath, the most fearful of Philistine warriors.

Creating the state of Israel required more than victory by slingshot. At the time Israel didn't exist. It was very much like Egypt before its sovereign ascendancy, which was divided into north and south territories. The people who would eventually form the state of Israel consisted initially of twelve tribes not unified politically. While David showed his military prowess in slaying the giant, it was his political finesse that sealed his reputation. In an assembly at Heron, he spoke to both the tribals and to his God, stating emphatically; (Samuel 5:1) Oh, *Lord they are all my bone and flesh.*

It was perhaps one of the first times a Jewish leader used a metaphorical reference to the body to symbolize tribal unity. It would not be the last. David's Nazarene descendant would later repeat the phrase centuries later in a different context, in the hope that Israel and its God would exercise his rule over the earth.

While David's success was unquestionably significant to Jewish history it also brought back to mind Samuel's warning about the rule of kings. David was a complicated man. He was said to sit at the right hand of God, yet he was also involved in intrigue throughout his tenure.

Though King David's religious legacy is mentioned prominently in the Old Testament, specifically The Book of Psalms and Samuel, his rise to prominence was the result of two factors: his political skill and his military prowess. His success in battle aroused jealousy in Saul, who set in motion a plan to assassinate David. This episode might have marked the point in time in which the Israelis finally fulfilled Moses' prophecy.

Moses knew anointing a king could lead to greed, cruelty, self-serving, expanded egos and disregard for the people. Some of that was occurring. It was up to David to solve the monotheistic/monarchic puzzle, to apportion humility, pride, passivity, aggression, populism, and the needs of the state. The fact that he became both religiously and politically important suggests he passed the test.

Not that he didn't waver. It was forbidden for Jews to marry outside their ethnicity. It was against the commandments to kill, especially members of one's own social group, i.e. his "brothers." It was also considered sinful to commit adultery. David's misdeeds extended to all these areas.

The tribes of Israel had been loosely constructed and without core leadership. They have been described as bands of outlaws living an opportunistic existence prior to David's intervention. Perhaps realizing this, David took the wives of some of these groups for himself and sired children by them. It might have been a way of centralizing the nation under one "father" who was both carnal and spiritual.

There was considerable dissent within his own family, possibly because familial distractions kept him from balancing the roles of king, husband, and father. Some accounts have him dispatching his son Absalom, who behaved in defiant manner and might have posed a threat to David's power.

Yet David ultimately succeeded, most notably due to his victory over the Philistines, which brought peace to Israel and allowed its people to expand their culture, production, and territory. At last, there was an actual nation and a "land of milk and honey."

Interestingly, David seemed to have more difficulty dealing with his own people than with foreign enemies. His relationship with Bathsheba was considered blasphemous by many Israelites. She was a Hittite and David impregnated her. That raised the possibility of the fledgling Jewish nation being contaminated by a foreign ruler at a future time.

Because David maintained stability within the Jewish state, the Lord chose not to condemn him, in fact, continued to hold him in high esteem - perhaps realizing there would be no reason to issue a set of laws to a group of chosen people who weren't galvanized in the first place. If Philistines continued to annihilate the tribes of Israel, took their women, fathered their children, and diversified the group, Jewry would have been extinguished, along with the God who was supposedly their sponsor. In a strange sort of way, God seemed to come to grips with the fact that he needed David as much as David needed him.

In view of such a pragmatic shift by the Lord, David might have felt justified combining the status of king and God, as had leaders in other nations. That might have marked the end of the Jewish moral theory but for two other gestures.

Ultimately, David retained the tribal-familial mindset despite his nation- building ventures. He knew it would be important to honor God by building a temple and that such an act would symbolize, more than anything else, the rise of Israel. Yet he harbored guilt, due to his awkward political and appetitive actions. As if through self-induced penance he deferred, leaving it to his son Solomon to build the magnificent temple and bask in its glory.

Another holy gesture came through David's pledge to God even at the height of his power. Samuel 6:1-19 refers to David's obedience to the Lord after being told to carry the Arc of the Covenant to Jerusalem. It was a clear illustration of submission, not so much for obeying God (he had no choice because God was adamant about this, even slew a man called Huzzah for accidentally tipping over the arc on the way). A more concrete indicator of David's loyalty to God was David's reaction upon entering Jerusalem with his precious cargo. He danced wildly, sang, recited poems, and acted as though in a trance. Finally, the musician, the poet, the gentle artist went back to his psychological roots. It seemed he was relieved to have the weight of governance removed from his shoulders by God.

David had re-discovered himself. He came, he saw, he conquered, he united an unruly group of disinterested tribals who feared the Philistines more than they loved the prospective nation of Israel. Sometimes he used charm, sometimes he used force, but he won an ultimate victory for the Lord.

Because of that, he was invited to sit by God's side, waiting patiently for centuries until a distant relative came along, at a different time in a vastly different political climate to re-invigorate beloved Israel. However, despite being at God's side, looking down, David probably knew the approach by the newcomer would have to be different - even King David couldn't have defeated the Romans.

The newcomer chose not to attack militarily. Instead, he utilized the weapons of compassion, love, and a precise dialectic to unite diverse peoples into much more than nationhood. He sought to create a universal entity much larger and holier than a tribal conglomerate. His quest was to create a state of mind, of the soul and of the heart without geographic, familial, or ideological boundaries. The man named Yeshua was a dreamer, whose moral universalism conquered more territory than Alexander – without lifting a hand.

As for David, in the end he avoided falling into the trap of arrogance. It was no small feat. Since the inception of nation states it has been typical for leaders to make themselves into gods. It began in Egypt but went far beyond the dynasties. Indeed, it seemed to permeate all cultures. Japanese emperors referred to themselves as gods, known as 'Shintoists'. Chinese emperors were called "Sons of Heaven". The Romans took this to an extreme degree. Tiberius and Caligula referred to themselves as gods. Constantine, who laid the foundations for Christian worship in Rome, ironically referred to himself as God. So did Hadrian and Octavian (the latter of whom renamed himself twice: first "Augustus"... meaning, "Most Venerable" and then "Divi Filius"... meaning "Son of the Divine One."

The trend was not limited to Europe and Asia. The Incas also considered their emperors to be gods, and even during the Renaissance, when, despite the humility-driven zeitgeist of Christianity, there was an associative fusion between man and God via the doctrine of Divine Right.

Despite the influence of Judeo-Christianity around the world such trends were repeated throughout history, making it appear as much representative of human nature as of politics or religion. In a socio-biological context, that is understandable. However, it also shows how the not-so-wise upright walker's own formidable mind can serve as his worst enemy. The alpha male acts in ways that reinforce his status, and his status reinforces his subsequent actions. The constant hormonal rush that results from that reciprocity will snowball over time, so that once human rulers decided they exist on a higher plane than their fellow humans, tyranny, and social entropy results.

To their credit, the leaders of Israel and David in particular, paved the way for an alternative approach, whereby the ruler would always defer to a higher power. Since that power was ultimately the God of all people it created a populist imprint on politics, religion and solidified the idea that people living in free societies derive their basic rights from God, not man.

CHAPTER 9:
THE MANY AND THE FEW

FOR ALL THE COMPLEXITY, TWISTS AND TURNS in human social evolution, it seems the historical process can be narrowed down to a dual process, pitting the needs of the individual (or the few) against those of the many. Homo sapiens might be described as a hyper-social animal, obsessively concerned with the actions, ideas, flaws, talents, and motivations of others. To a large extent, the reasons we are so socially inclined have to do with the layout of the human brain. We have language centers in the frontoparietal cortex that regulate the movements of lips, tongue and larynx and facilitate sound making. In that same area we have association centers which enable us to pair up objects, people, and circumstances in terms of both their distinct and common features. That means we can use our sound making skills to develop labels and connective associations between and among a wide variety of events, objects, and people.

We also have what is called a tertiary associative cortex, which enables us to think in representative, abstract and hypothetical terms. Finally, our exceptional memory capacity allows us to keep others in mind even after death, which leads to myths, heroes, villains, and gods.

Due to a capacity to super-categorize and integrate our social world, and because we can say, feel and think so many things about our fellows, the social world consumes a a great amount of the human vista. In simple terms – as was discussed earlier - having a large brain guarantees strong social interest.

Yet there is another aspect to this which pertains to individuation. The same neural hardware that enables us to super-categorize people enables us to think about ourselves as separate entities. We can post-process our actions, not only through feedback from others but also from our internal value systems in determining if what we do is consistent with what we know about ourselves.

In other words, we have what no other animal has - a sense of self. While Freud discussed this in his description of the ego, it is more closely related to what psychotherapist Carl Rogers called the self-concept, and what Harry Stack Sullivan called the self-system. Being aware of ourselves - as though observing another, means that during a lifetime each of us will store an enormous amount of information regarding our needs, aversions, feelings, hopes and dreams. Over time, that will tend to dominate our mindset and guide our reactions.

That doesn't mean we're selfish - most of us are not. It does mean that we will always tend to be more aware of our status, needs and wants than of others. Almost every social perception emanates from the self, which functions as an existential pivot point. In that context, the classic conception of projection as a defense mechanism might not be entirely valid. For example, while projection - the tendency to attribute one's traits to others, is described as a defense mechanism, (which suggests a distortion of reality) it, might well a neuro-behavioral imperative.

Our achievements are typically evaluated in comparison with others. We exist in terms of norms, which means our lives are both individualistic and collective. Our test scores in school, our jobs, our health status, our reactions to positive and negative experiences are all dealt with psychologically in socially comparative terms.

Interestingly, having such an ingrained sense of self enables us to feel empathy, because in doing so we are responding to others' pain as if we had the same experience. In fact, we don't so much empathize with others as project an image off ourselves onto them to zero in on their experience. As odd as it sounds, emotionally and perceptually, humans operate a bit like fruit bats, who utilize echolocation to determine the layout of their surroundings. They do so by sending out feed forward, rickety sounds that bounce off objects in their path, enabling them to process thefeedback and navigate effectively.

Human social perception is analogous to that because most of what we perceive reflects what we are. Cruel people will tend to interpret the motives of others as being cruel. Affectionate types will tend to be more trusting. All of that pertains to the self-system but can also be applied to human history, particularly regarding behavior patterns with the advent of urban, densely populated settlements.

In that context, a question comes to mind. Which is more important, individual rights or group cohesion? That question has been raised in every political theory since man first began recording history. Pharaohs like Cheops, Setti and Ramses emphasized the needs of the individual over the group. Machiavelli did the same, although he somehow managed to equate the whims of the ruler with the needs of thepeople.

Philosopher Jean Jacques Rousseau did as well, with his concepts of natural man, the social contract, and his contention that the individual is corrupted by society rather than from within. Later, Thomas Jefferson and James Madison reinforced the notion of individuation by conjuring up a Declaration and Constitution that made individual rights so predominant that the American legal system favored protection of the innocent from wrongful imprisonment over the possibility of the guilty going free.

The emphasis on individuality has come under fire throughout history, most notably when the distribution of resources seemed so skewed toward the few as to foster rebellion. All socialist and communist systems have used economic duress as a reason to shift from the needs of the individual to those of the collective. That shift has typically led to oppression and disregard for the daily lives of the populace (whose experience, after all, is always based on individual occurrences and interactions).

The individual vs. group conundrum persists to this day in the conservative vs liberal ideologies. The argument has never really been settled, which is why no truly comprehensive political system has ever been devised. Ultimately, the solution requires a sense of proportion that involves, taking both individual freedoms and the common good into account.

Unfortunately, ideologues find that discomforting. Regardless of the political discourse, differing ideologies and party loyalties, almost all political arguments seem to revolve around the search for what might be called (in lieu of Utopia) an ideal sociopolitical proportion.

In many ways, history has been typified by ideological shifts back and forth between individual freedom and the common good. It seems, when one or the other system leaves people wanting, they drift, almost reflexively toward the opposite ideology. It might be the group/individual conundrum captures man's entire journey through history, and one that arguably, we might not be smart enough to resolve.

Each system has clear goals and outcomes. One type of system, socialism, is group oriented. The idea is to dispense wealth as evenly as possible to create some sort of egalitarian climate. It does not completely obliterate rewards earned through individual achievement but comes close. Communism does preclude individuation, especially in economic terms. There is no private property, as per Marx and Engels' tenet...*from each according to his efforts, to each according to his needs.* That has proven disastrous; not just in political terms but because it denies two core aspects of human nature: first, by precluding individuation and the establishment of identity, and second by stifling the motivation that goes along with the expectation of receiving rewards for one's behavior.

Conversely, capitalism favors and encourages individual achievement, and its typical correlates, wealth, and notoriety. It works well for the most part, due to the resonance of consumer activity. Because merchants will always need laborers to make a profit, wealth tends to increase through hiring practices, investment, and charitable donations.

Capitalism also drawbacks. One is resentment among the have-nots, especially in societies where equality before the law is interpreted more broadly to mean equality of outcomes. Another drawback is that in encouraging people to set and attain goals, it suggests success is always within the grasp of the motivated. That, of course, is not true. Thousands of highly motivated restaurateurs go out of business yearly due to factors unrelated to their efforts. In fact, capitalism is competitive, and each race has winners and losers.

Still, capitalism seems to win the proportion argument. It produces more economic growth and, if well-regulated by unions, antitrust laws and competition leads to policies that have a broad, beneficial effect on people's lives. Arguably, the most important byproduct of capitalism is not its direct economic impact but its allowance for creativity. It encourages brilliant, highly motivated minds to innovate, solve problems and move society forward.

Another argument in support of capitalism is that (without completely dismissing the importance of "group think" and social conformity) the progress of human society has been orchestrated largely by individuals. Alexander Graham Bell invented the telephone. Jonas Salk developed the polio vaccine. Einstein discovered Special and General Relativity, Newton figured out the mathematical nuts and bolts of gravity and Freud changed the ways in which we think about human nature. Each of these talented individuals had mentors. Freud had his

Mesmer, Einstein had his Maxwell. However, the boldness and creative innovation in these instances ultimately arose from single minds.

What about the group? The main benefits of a large group philosophy, especially during the early civilizations, were territorial defense and production. Greater numbers produce more products and greater numbers tend to be victorious in war. That is perhaps why there has always been a strong correlation between a nation's imperialistic intentions and its social policy. Hitler needed a large army to support his grandiose ambitions, thus created the National Socialist Party and recruited hordes of German youth into the fold. Lenin wanted to spread communism around the globe. His recruiting policy targeted not only young people for the military but "workers of the world."

There are, of course, other benefits of the group orientation. In large groups, there will be a vast exchange of ideas and there will be competition to see whose ideas, inventions and policies work best. In his book, Wealth of Nations, Adam Smith described a moral anchor point within the capitalist system as deriving from competition. He assumed that when companies vie for consumer dollars, prices will tend to drop, products will tend to improve, and no single economic Leviathan will emerge to control the marketplace.

The group vs individuation question has also played a role in religious history. Gods often had to juggle the two considerations. The God of the Hebrews usually acted in terms of the greater good. His punishments could be harsh, but his actions sent a message he hoped would be heeded and lead to group cohesion. With group solidarity he seemed to feel the nation of Israel would thrive and expand. His acts were typically group oriented.

On the other hand, by tolerating David's individual quirks, he demonstrated that proportion, as well as ideology, matters. David understood this, as he demonstrated by re-affirming his allegiance to God in the end.

It would be fascinating to go back in time to see how this played out in evolutionary terms. In the early human tribes, the contributions of all individuals were likely considered valuable. The need for workers, coupled with small numbers probably obviated to an extent the need for social rank. Surely there would have been leaders and followers, but in a group consisting of roughly 150-200 members, a climate of tyranny would have been counterproductive.

Also, in a small, highly functional group, mating practices might have been liberal. The necessity of producing offspring to continue the line would have fostered heightened sexual interest. One byproduct might have been an egalitarian, liberal sexual ethos. That would have interfered with the typical primate mating practice that involved passing superior (alpha male) genetic traits onto the next generation, but at the same time would have provided the benefit of pan-social investment in all offspring. In other words, …every child is *our* child.

In that context, it is interesting to note that archeologists usually categorize the skills of hominids as a group, as if all individuals had the same intellect and talents. For example, it is assumed that Homo habilis was the first to make tools. It is also assumed Homo erectus was the first to discover how to make and use fire, and that Neanderthal figured out how to contain fire in makeshift hearths. However, it is possible that individual in those groups used their creativity to invent and produce these utilities while the rest of the group copied from them.

It seems nature has arranged it so both the individual and group factors are crucial to human survival. To the extent that our genes influence behaviors the formula for socio-biological survival and adaptation seems to involve a capacity to think and act proportionately, in accord with the needs of both the individual (or the few) and themany.

Religious figures throughout history have grappled with that duality. Each approached the problem differently.

Siddhartha Gautama (the Buddha) began his life in a highly individual context. He was born in Nepal to a family of great wealth and was himself slated to become a prince. Then came a complete reversal. He ultimately rejected the perks of royalty and instead embarked on a mission that was so broad in scope as to encompass all of nature. His teachings were old and new. He was both a religious innovator and a figure espousing a theology reminiscentof pantheism.

In a sense, he was also both a revolutionary and a throwback. He could be considered the originator of a pantheistic revision now known as humanism. He believed man was at one with nature, that the lives of all creatures were valuable, and that in order to be truly moral, one had to look within to find the peace that comes with a humble retreat to nature's holistic tapestry. His focus was on the alleviation of human suffering. His new religious model was both religious and introspective. He offered only four commandments which could be more accurately described as moral tasks.

To find the truth of suffering
To find the cause of suffering
To find the truth of the end of suffering
To find the truth of the prevention of suffering

The Buddha's view of mankind was unique. His version of pantheism included more than "oneness." It included empathy and cures for the maladies of all of nature's creatures. His universality emerged in the face of extreme parochialism – it was revolutionary.

In early tribal groups empathy would have been confined to relatives and fellow tribesmen because that's all there were in those enclaves. But this trend carried on. In early religions and social systems, providing help and aid to strangers was, if not forbidden, then certainly considered anomalous. This ancient ethic likely reflected a biological trend toward genetic localism. It was not wrong. It was just atavistic.

While people in modern times automatically equate the term 'genetic purity' with racism and fascism, there is a tendency among all organisms to favor the core genetic family unit in their actions. Humans can override that with linguistic concepts, enabling us to create connections among disparate populations. However, prior to the advent of integrative beliefs the need to sustain the family line was fervent and our evolutionary proclivities once had a stronger grip on how our predecessors felt and acted.

In the New Testament, this is exemplified in Luke 10:25-37 which tells the story of the Good Samaritan. A man in desperate need is ignored by a Jewish priest and a Levite while assistance is provided by a Samaritan. This is described by Luke as admirable but odd behavior. Luke clearly had a strong moral foundation, but he was also aware of the Zeitgeist during his lifetime. There were several reasons why compassion for strangers was atypical.

The Ice Age on earth ended about 12,000 years ago. Not long after that came a global thaw which released water to the plains. Warm winds carried seeds more prolifically around the plains of the Mediterranean and Middle East. Over time grains and other edible plants began to grow freely and robustly. Once the process by which these plants grew was discovered, agricultural settlements cropped up. This attracted nomadic tribes and led to centralization of increasingly dense populations. The transition from tribal life, which involved small, manageable populations that traveled freely, might have been beneficial by making food more available. However, it also effected the human/nomadic psyche. Moving into congested urban areas, where travel was restricted, where there was a strange phenomenon known as property ownership, and where rules of behavior were enforced by remote, central authorities was new to them.

Such restrictive trends arose out of necessity but only because the trustful altruism previously applied to relatives did not apply to people from other families and tribes. In other words, two byproducts of the agricultural revolution were alienation and an increase in antisocial behavior. For a time, well-fed, but confused Homo sapiens was out of his element. Stranger anxiety was intensified, leading to compensatory attempts to preserve the mores and identity of the tribe.

The tendency toward small group, familial tribalism had prevailed for millennia and was reinforced by both evolution and social conditioning. Consequently, it would have taken very special people to overcome socio-genetically driven parochialism and build a bridge of compassion linking all humanity. Very few had the vision to accomplish that.

That was Buddha's quest. He was not immediately accepted. His universality, like that of the Samaritan, was considered odd but his imprint on humanity was distinct and ultimately indelible.

With the advent of a new humanist movement the clouds of confusion were lifting. A partial solution to the group-individual conundrum was at hand. Someone had taken on the task of overriding biology through faith.

Buddha wasn't merely challenging parochialism. He was also renouncing the tendency among early religions to reward mere obedience. His new concept created its own evolutionary trend, featuring a compassionate, inclusive approach free of parochialism. It might have been the first time in history that a significant person conceived of all humans as belonging to the family of man.

The question was whether Buddhism could override human psychobiology. Certainly, the scope of his faith went well beyond tribal parameters. On the other hand, the interpersonal closeness implied in his teachings was arguably a replication of the same sentiment typical of the tribal mindset. In that sense, the Buddha can be said to be recombining, rather than replacing an inborn human social trait. He was not a man intent on breaking down a preexisting system but a man/ deity who wanted to insert the naturalism and social cohesion of the tribe into the cold detachment of the nation state.

The idea of compassion as a religious default position was new and invited scorn in some quarters, but it was attractive to some, because its logic was impeccable. Since God views birds, bees, and all creatures in benevolent terms, he would surely do the same with even the lowest human being.

As a significant number of people became aware of this movement, a wave swept over the world. People brushed aside in previous times saw the light of inclusion. Though politically irrelevant, the poor were not only invited to the party but empowered. They did not have to petition or cave to a tyrant. To the contrary, Buddhism proved control could be gained internally, that heaven could be attained within mind. The people were the power, the arbiter, in a sense both human and divine. As a result, people did not have to fear God's wrath. Because He was within them. Nor was it necessary to sacrifice lambs. According to the tenets of this new faith they simply had to meditate.

It was a first step toward ending an era of punitive, submissive religiosity. Now, man could observe his faith through self-consultations as well as reverence for God. He could attain a mental state called Nirvana quite on his own.

While other religious figures undoubtedly conjured up similar ideas, it was the Buddha who spread the word most effectively. His actions created a new social outlook with altruism as the prime doctrine. Not surprisingly others followed.

It is not known how far The Buddha's travels took him. Some accounts suggest he made his way to the Middle East. If so, he could have imparted his beliefs several centuries before another humanist came along. Whether there was a Buddhist influence on the new guy is hard to say. There are no historical records suggesting as much. Perhaps that was irrelevant.

Sometimes ideas seep through without necessarily emanating from a single source. Fact is a new doctrine was in play. It was time for religion to turn inward, from faith based on obedience to faith based on love. A point man from the town of Nazareth lay in wait.

CHAPTER 10:
THE REBEL MESSIAH

WILHELM GEORG HEGEL WAS A GERMAN PHILOSOPHER who believed history unfolded deterministically. He did not claim to be able to predict specific times and events. Rather, his model suggested changes in social structures were inevitable and that this process occurred in stages. He called the first stage *thesis*, which referred to the initial existence of a particular government format. The second stage was called *antithesis*. This referred to the fact that all governmental systems run their course, then are overturned, either through revolution or cultural decay. The third stage was called *synthesis*, and this referred to the fact that at some point a better, more progressive society would emerge from the rubble of the previously flawed political system.

While Hegel presented this as a novel theory of history it was likely influenced by Plato's belief that all things in nature evolve into perfect forms. The difference was that Hegel viewed this transition in a spiritual context. He assumed a transcendent process or entity would orchestrate the change. Whether he realized this idea would eventually serve as justification for the rise of communism is not clear.

Since Hegel was a spiritual man and communism prohibits religion, it appears Marx and Engels co-opted the model to suit their own purposes However, this evolutionary concept of history created all kinds of possibilities.

Hegelian theory is no longer considered relevant, in part due to the failure of communist societies. However, the notion that history follows certain sequential rules persists in some quarters. Over time, theories on the evolution of human society have revolved around a few main themes. It probably began in Greece.

Historians Herodotus and Polybius believed the history of human society unfolded in accord with changes in human behavior. They felt If mankind waxed cruel and dismissive during a particular epoch, the ideas, politics, and culture would follow from that, including religious practices. Other philosophers like Hippocrates (who was much more than a physician) and adherents to the Atomist school believed that while the historical process changes with time it simply evolves in terms of core,unalterable aspects of human nature.

Theories of history that have included existential as well as economic forces in the cause - effect process. Theorists and historians have attempted to describe causal trends between the time of the first Mesopotamian settlements and the current time. Many of these conceptions have provided insights into human nature. Some were accurate, some not. I believe the most accurate account of history was provided by Sigmund Freud. In *Civilization and its Discontents,* he proposed that the first human beings were initially id dominated. Like other animals, they hunted, competed for mates and territory, expressed their sexuality freely and did not self-evaluate enough to foster anxiety and guilt. Freud believed it was only after heavily populated settlements arose that the restrictions and laws were needed to maintain order. At that point, the inhibitory circuits of the human brain were effectively re-structured to incorporate the egoand superego.

The human brain includes a midbrain circuit collectively referred to as the limbic system. Despite its concern with prime survival and appetitive functions, it has vast connections with neocortical circuits devoted to language, self-perception, and anticipatory cognition. Because cortico-limbic interaction requires on-going compromise between urge and reason, the psychic primal-moral juggling act described by Freud might more accurately describe the ways in which human nature has shaped the course of history.

Freud's theory suggests the formation of the personality paralleled the journey of man from the wild to the city, that the politics and laws in ancient urban cultures operated on three levels. They doled out resources in some reasonable proportion to meet the needs of the people. They reined in the atavistic side of Homo sapiens, and they attempted to strike a bargain between individual/small group concerns and the organizational needs of larger society.

In a sense, the challenge for all religious figures has also centered on those factors. The early Christians resented having to pay taxes to Rome, a civic entity with whom they had no political, emotional, or religious connection. They also resented being punished merely for being a distinct religious group and for acting according to their own beliefs.

Although his mission was spiritual, Jesus of Nazareth had to factor in those concerns during his three-year ministry. His appearance on the scene was curious. The moral revolution already had begun in Judea through the efforts of, among others, John the Baptist. At the time of Jesus' baptism in the Jordan River it was not clear John would be executed. He was controversial. He did confront Herod Antipas, but Herod was also a Jew who feared the Lord's wrath, thus paid heed to John's admonitions.

If John was still relevant at the time of Jesus' baptism, the question is, why the transfer of power from John to Jesus? Earlier prophets, particularly Isaiah in 7:14 had predicted the coming of a messiah, albeit in general terms, that is, as a descendant of David. Theoretically any figure who claimed to be descended from David could have stepped forward, including John, who was related to Jesus and therefore at least indirectly in the Davidian line.

It is curious that Luke 2:39- (depicting the birth scene in Bethlehem) mentioned that Jesus's family was from Nazareth. Yet Bethlehem was a considerable distance away. The New Testament does not provide a definitive reason for his birth in Bethlehem. It is stated in Micah 5:2 that a messianic era would begin in that town. However, descriptions of the man who would assume that role were vague. Micah's passage is as follows:

And you Bethlehem Jephthah, you being among thousands of Judah out of you He shall come forth (be born) to Me. That is to become ruler of Israel.

Many biblical scholars believe this referred to Jesus. However, that seems questionable. Back then, royal status was typically passed down through generations. The first successful king of Israel was David, and most assume Jesus' birth in Bethlehem was a step toward proving a connection between David's line and that of Joseph and Mary. That would have given young Jesus a foothold in his claim to the throne. Then again, how is it that a descendant of perhaps the most illustrious figure in the Old Testament ended up being fathered by a poor carpenter?

The Davidian model seems odd for other reasons. David was a musician - Jesus was not. David was a skilled warrior who killed a fair amount of people and arranged the death of one of his generals. He took many wives. He seduced (or was seduced by) Bathsheba. David was a skilled, calculating politician who established himself by defeating the Philistines' formidable army.

Conversely, Jesus seems not to have had a political bone in his body. He never even hinted at forming an army and, of course, the Roman legions were infinitely better trained, better supplied and more ruthless than the Philistines.

While it is hard to figure out why the Davidian model was adopted there is one element that could have created a crucial connection. It was a statement made by David after being sent by

God to the town of Heron. A long war was taking place among the tribes. Saul and David were, at this point, leaders of two separate factions. It appears God wanted David to go to a place relatively free of conflict and invite the twelve tribes to convene. At that time, it was clear that for Israel to prevail required formation of a kingdom. Eventually God's purpose became evident when, as referenced in 2 Samuel 5, David stood before the throngs and stated: "You are my bone and my flesh and blood."

By poignantly citing the ties shared among the tribes - all of whom believed they were descended from Jacob - David became the great unifier. In that context, Jesus' connection to David makes sense. He was charged with the task of unifying Israel under one God, as David had done.

However, interestingly, Jesus seemed to be seeking a more global unification. Indeed, his universal inclusion of all mankind was more reminiscent of The Buddha than of prior Jewish kings or prophets. Jesus was not merely uniting twelve tribes, although there was enough discord among Jewish sects back then that it would have been a worthy pursuit. Somehow, some way, he believed he could bring together the entire family of man. While this strategy might have seemed odd to Jews at the time, it could be Jesus was such a doctrinal purist that he took the words of Isaiah literally.

In 11:10 the prophet said: "The root of Jesse shall stand for the ensign of the people; to it shall the Gentiles seek." So, while Roman oppression forced Jews into a primary focus on their own customs and faith, Jesus was likely working on a broader theosophy thatwould draw in even Gentiles.

Almost 500 years after Buddha's journey the humanist doctrine was once again in play. Now there were two theosophical rebels espousing the idea of a supra-biological belief system, based on 200,000 years of nomadic, tribal imprinting.

This new movement espoused by Buddha and Jesus was destined to grow because man is what he is. While the flexible human brain enables us to adapt to varying circumstances and environments, there are fundamental behavioral and emotional traits that will always prevail despite political, cultural and social changes.

Among them is the small tribal social ideal. Our species chose to adopt an agrarian lifestyle. The earth's warming period led to the availability of edible plants. That, along with the domestication of animals created opportunities for small homo-indigenous tribes to shift from nomadic to urban lifestyles. However, expansion into heavily populated centers was not driven only by agriculture. Nomads could have sustained their preferred population and still cultivated the land. Instead, it was war and the quest for ever-expanding territory that set human society on its ear.

Primates are territorial. Chimpanzees are our closest genetic relatives, and they are constantly fending off invaders from other groups. Invasions are typically characterized by the same kinds of behavior seen with humans, including the slaughter of opponents, capturing, and engaging in forced sex with foreign females and murdering the young. While chimps are not adorned in Viking regalia, do not paint their faces before battle, and cannot appeal to Mars or Odin for success in battle, the behavioral mechanics are essentially the same.

Chimps deal with this problem case by case. They are bright creatures, but they do not have a large enough brain to draw up intricate plans, anticipate outcomes or express themselves persuasively enough to form bonds with strangers for purposes of territorial defense. Their phonetic alphabet consists of roughly 19 or 20 distinct sounds, all of which signal imminent events such as the presence of enemies, snakes, or leopards. They have little sense of the past or future. Obviously, humans do, and one byproduct of that capability seems to have been psychological tolerance for the expansion and integration of diverse populations.

Another common primate trait is seen in the importance of the alpha male. In times of war, leadership is needed. The people of all civilizations realized this, including the Israelites. Saul's foray into kingship and David's ascendancy were the result of God's assessment of the times. He knew war would be a constant concern, that preservation of the people he had chosen required adequate defense of territory. In fact, He was Himself rather imperialistic in urging the Israelites to move into lands occupied by other peoples.

War changed everything. It formed attitudes, religious beliefs, and practices, created cultures, produced hybrid philosophies and languages and propelled the human race into modernity.

However, it was also horrific. The construction of roads, weapons, the training of soldiers, horses and building of temples came at a cost. The cost was bloodshed and loss of tribal identity. In effect, war was not just a conflict between opponents. It was also an assault on the biological mandate of keeping intact the core genetic line. Cooperation among strangers became a useful trend that nonetheless upset the existential apple cart of our species.

Any intelligent person observing such trends during the Common Era would realize humanity would never again return to the nomadic, tribal lifestyle. Nation states were here to stay. Therefore, the prime challenge of prophets and politicians was to work within that context and find some way to stop the madness.

That is one of the least recognized contributions of Jesus Christ. Whether viewed as a god or higher-level prophet like Mohamed, he was a quintessential rebel whose philosophy differed so much from those around him that he came across as a confusing figure, even to some of his followers. Movies about his life tend to emphasize miracles as the main drawing card of his ministry, but to this day it is sometimes difficult to discern his motives. Perhaps the sheer magnitude of his aspirations and eventual success in creating a movement that spread all over the world (including in Rome) makes his presence on earth and his ultimate impact so grand as to be incomprehensible.

Obviously, the stated goal of his mission was to offer himself up as a sacrifice. In so doing, he would absolve man of all his sins and open the gates of heaven. Yet, there seems to have been so much more to his belief system.

He was, of course a Jew, well versed in the Torah and the details of prophecy. He was guided by the words of Isaiah, who predicted the arrival of a meek, yet strong messiah - a suffering servant, able to change the world yet, ironically, unable (or disinclined) to save himself.

However, the New Testament makes it clear that self-sacrifice was only part of his mission. This is illustrated in Luke: 36. which refers to the statement by a listener about "this amazing doctrine" followed by the statement... "what a word is this."

The "word" reference has great significance. The language Jesus used would have been consistent with prophecy, but the reaction of the listener raises questions. On one hand a person with little knowledge of the intricacies of prophecy wouldn't be awed by Jesus' interpretation. He wouldn't necessarily know enough about it to experience confusion. Conversely, if versed in the tenets of Judaism, a listener wouldn't necessarily find Jesus's language so amazing. It presents a bit of aquandary.

Several possible answers come to mind. The "word" comment could refer to Jesus' unique gift of persuasion. At the onset of the Common Era Israel was quite culturally diverse. Jews were, perhaps begrudgingly, influenced by Roman culture. Romans transformed every society they conquered through the impact of the Latin language, Roman law and especially Rome's ingenious methods of civil engineering. Yet Rome was influenced by the Greeks, which meant their influence on subjugated cultures was really Greco - Roman. That is significant.

Alexander's troops had traveled to that part of the world several centuries earlier. He was more than a military leader, indeed brought the ideas of intellectuals, artists and scientists into whatever territory he subdued. Thus, due to the dual and powerful draw of both Roman and Greek culture the entire Middle East had experienced substantial transformations prior to the arrival of the Romans or the birth of Jesus.

One of the influences would have been the dialectic method, as developed by Socrates and later employed in Plato's Academy and Aristotle's Lyceum. This was an educational teaching tool in which, rather than dogma and straight-line memorization, students were required to resolve conflicting points and work toward a resolution. It was essentially a format based on syllogistic reasoning - what is referred to in modern times as logic, or the dialectic.

There are numerous examples of Jesus' use of the dialectic in the New Testament. For example, during interrogation by the Pharisees (Luke 20:4) Jesus was asked where he got his authority to preach.

Rather than issuing a direct answer, he drew a comparison between himself and John the Baptist, asserting that if John could preach the word despite not being an official of the Temple, why not Jesus himself? He used the same comparison method in responding to the San Hedren about the similarity between his and King David's relationship to God (Matthew 22:41-46). The tactic of answering a question with a question was classic dialecticmethodology.

Other quotes suggest a Hellenic/dialectic influence. For example, In Luke 6:29, Jesus rendered his now famous statement on "turning the other cheek." Considering the eye for an eye, tooth for a tooth maxim seen in Exodus 20:24 that was a bit of a hairpin turn but there is a possible Greco - Roman source. The same phrase is presented in Plato's seminal work, *The Crito*.

Another Hellenic-Christian connection is seen in the similarity between Plato's comment: *A house united cannot be defeated,* and Jesus' statement in Mark 4:25: *A house divided against itself cannot stand.*

Still, another connection is seen in Jesus' emphasis on life on earth. His concept of a kingdom was not like David's. It was in heaven, along the lines of the Greek belief in Elysian Fields. It seemed to deviate from basic Judaic doctrine.

Pagan religions typically adopted the belief that the privileged would gain posthumous access to paradise. Most Jewish sects acknowledged God's place in paradise but did not emphasize this in their doctrine. Jews believed the reward for religious adherence was here on earth. Jesus did believe in heavenly reward. Indeed, it was the prime reason behind his mission and for serving himself up as a sacrificial lamb.

Many believed in an afterlife back then, but paradise was often viewed as an exclusive abode accessible only to a few. Jesus, on the other hand, came along and opened the gates of heaven to all. It is impossible to gauge the impact this would have had on commoners and slaves back then. To them, existence consisted of living and dying. Life for a laborer was always hard. Pain and suffering were the norm. The monotony of the task, the grueling, non-stop requirements involved in construction projects, often leading to untreated injuries, added up to a futile existence in which rewards were scarce. The idea that hardship was a precursor to eternal life in paradise would have been profoundly attractive and persuasive. It would have taken persuasive, exquisitely logical language to create such a profound change in beliefs.

Within the Jewish community, this was a fairly new approach to teaching. Indeed, during Jesus's interrogation the San Hedren were so overwhelmed by Jesus' syllogistic elegance and his promise of heaven for all that they asked no further questions. While the decision was ultimately made to crucify Jesus, he was at least in this instance, winner of the debate.

The possibility that Jesus used the dialectic rather than emphasizing abject obedience to God is interesting. Some theorists believe the Christian doctrine derived from Hellenic (Greek) influence. Daniel Graham and James Tieback discussed that possibility in their book, *Philosophy and Early Christianity*. It makes sense, because the sermons of St. Paul and the writings of St. Thomas Aquinas and St. Augustine featured heavy use of the syllogism. Jesus' use of this method involved an interrogatory approach. Perhaps., like Socrates, he realized listeners become more engaged by questions than assertions.

There are other important features in this shift in tone. Perhaps the most important was deference to the thinker. It empowered people to reason their way to faith. It was egalitarian, not a question of a super authority lording over the people but of an interaction leading to resolution. More than the working of miracles, that might explain why a modest Nazarene, not well-received in his own hometown, eventually became so prominent around the world.

Whether one chooses to believe Jesus was a God or prophet/philosopher he was surely one of the great teachers in history. His sermons appealed to both the passion and the intellect. That made him difficult to resist. Over time, he captured the hearts and minds of so many around the world, even in Rome, where, despite governmental restrictions, ten to fifteen percent of Romans had converted to Christianity even before the reign of Constantine. In that context, it seems Constantine's Edict of Milan in 313 which legalized Christianity in Rome, was predictable.

Christianity had been seeping into the Roman mindset. Jesus was popular with the disenfranchised. Indeed, it wasn't so much the worship of Christ that led the Romans to persecute Christians in the first place but the latter's refusal to pay tribute tothe Roman gods.

Another reason for the spread of Christianity was because its adherents had the advantage of numbers. It wasn't just the poor who were invited into the afterlife. It was also the sick, the despondent and even sinners who sought forgiveness. That was incredibly attractive relative to what was offered in other faiths.

Pagan and pantheistic religions are functional. The gods provide crops, victory, fertility, rain, and other necessities. The pagan God is a provider but he or she couldn't necessarily look inside the minds of worshipers to feel and resolve their pain. The God, as redefined by Jesus was also interested in providing comfort in times of duress. One did not have to win a battle or build a temple to get His attention. Rather than being a God of winners, he was a God of dutiful participants. All one had to do was engage, believe, and have faith.

The appeal of such a doctrine would have been enormous. Jesus was effectively a populist Messiah. He did not necessarily encompass all of nature's creatures under his umbrella of empathy as did Buddha, but his announcement to the world that virtually everyone was welcome to the party in a time of extreme divisions based on class and heritage was about as seductive as it got.

In this book, each religious epoch and theosophical transition has been discussed in evolutionary terms. In that context, it seems fair to ask why a dialectic method, addressing the internal needs of people, would have caught on and spread worldwide? In other words, is there some sort of neuro-behavioral factor that makes that method attractive?

It turns out Buddha, Jesus, and the ancient Greeks were on to something. The human brain evolved to deal with the pressures of multiple environments. To succeed and prevail required brain circuits without fixed function that could form novel associations and enable our ancestors to invent, explore and conceptualize. Such a brain would have to be neurologically open to deal with uncertain, unpredictable stimulus-response pairings. It would need to operate through an initial general arousal process (what neurologist Karl Lashley referred to as mass action) and narrow it down to a specific response or thought process. For that to occur, the brain would have to experience a pleasure response in the transition from confusion to resolution via what is now called the 'aha experience.'

The fact that the brain derives pleasure from reduction of conceptual conflict makes it perfectly suited to the dialectic method employed by Jesus and other great thinkers. The importance of this process in learning and persuasion is perhaps best captured in the previously referenced statement by Sir Francis Bacon regarding the pleasurable transition from uncertainty to resolution.

In that context, it seems the doctrine and methods of Jesus and subsequent theologians who adopted similar methods would have been consonant with human nature. Listening to Jesus, Plato and Aristotle was simply more entertaining than listening to dogma.

Another feature of the dialectic method is mnemonic enhancement. Because the listener participates in the discussion and must "work" his neurons in search of a solution means there will be more cognitive investment in the process. That enhances attention to the subject matter and the capacity to post-process the information. Because the message resonates it is more entrenched in mind.

In educational circles there is an old saying. An average instructor teaches to the curriculum, a good instructor teaches to the students. The shift in communicative style from dogma to dialectic is not typically considered historically important, but it was as revolutionary and progressive a tool as later occurred with the steam engine and the computer. It's emergence as a teaching method did not emanate from a single culture, religion or individual. The usual assumption is that it originated in Greece, then became widespread throughout the Mediterranean because Aristotle's pupil, Alexander introduced the method to people in conquered territories.

However, since the dialectic is brain-friendly, it was probably utilized before that. For example, Hammurabi employed a dialectic format in creating a proportionate system of laws. His code was awkwardly skewed toward favoring those with high status, but it was peppered with dialectics.

The dialectic was also used in Egypt, despite its devotion to mysticism and hybrid gods. In fact, the Egyptian education system likely influenced Greek thought. This was pointed out by Plato, who traveled to Egypt on various occasions and took notes on how children were taught in elementary schools. His account described their methods in a way reminiscent of what is now called the Montessori method. Their approach presented open-ended questions which children were asked to explore and resolve on their own. It was an active, interrogatory approach, very much in accord with human brain function. Whether this resulted from Greek influence or occurred separately is hard to tell and is probably a question of proportion.

In any event the method had circulated around the known world and eventually made its way into religious doctrine. Those most able to utilize it in their prophecies and sermons tended to be the most successful and historically relevant. In Judea, its most notable proponent was Jesus, but this was just one of his methodological innovations. In addition to the dialectic, Jesus utilized other communicative skills.

The climate in Judea was intense, angry, rebellious, and paranoid. Pilate's legions were not numerous and imposing enough to completely discourage revolt. Revolts occurred occasionally, albeit typically through single episodes, assassinations, or zealot uprisings. Jews lacked training, weaponry, horsemanship, and finances to defeat Rome. Therefore, the task for a spiritual teacher was to persuade without inciting.

As with Buddha, Jesus was operating on two levels. One had to do with the politics of religion, while the other derived from a more primal, homo-indigenous cognitive template that simply could not be resisted. Egalitarian compassion, the need to summon the efforts and commitment of all members of the group, to acknowledge emotions as well as practicalities were factors needed to sustain group cohesion. All these elements were probably typical of early nomadic social groups and of the new Christian doctrine proposed by Jesus of Nazareth.

In reviewing the ministries of Buddha and Jesus it seems Homo sapiens' social and cultural journey has been an occasionally interrupted but unavoidable trip back to a social model with built-in components of equality, compassion, maximal respect, and use of human potential.

That suggests the notion of social progress might be somewhat illusory, and that despite the neuro-behavioral plasticity of his brain, man is restrained by social instinct, or at least behavior patterns so important to survival that they could not be discarded by history, politics, or religion.

That raises a question. If all mankind can do in the areas of politics, law and religion is improvise on a compelling cognitive model tethered to the psychology of the tribe, how does one describe human progress?

Despite the theories of Kant, Hegel and Marx innate, relatively unalterable social schema might typify and limit the human socio-biological mindset. If so, then the primary impact of Jesus of Nazareth might be that, like the Buddha he did a hell of a job trying to fit a large society into a small one and in the process, lead us all back to the blissful necessity of humanism.

CHAPTER 11:

THE AFTERLIFE

ONE OF THE MOST BASIC, YET MYSTERIOUS ELEMENTS in all religions is a belief in eternal life. Many religions operate on a reward and punishment system. Pending judgment of his or her moral standing, an individual will either live forever in a place with no duress or need for material acquisitions, or in a place with permanent or (if sent to purgatory) temporary suffering.

While heaven and hell are aspects of the Abrahamic religions, descriptions of heaven and hell in all three faiths have been only ambiguously described. The ancient Hebrews addressed this subject reluctantly, perhaps due to reverence for a god they felt was so far above man that contemplating a posthumous co-existence with him would be blasphemous.

That changed in later times. By the time of Jesus' ministry there was a growing belief in an afterlife, accessible not just to exalted individuals but to all the faithful. This transition might have marked a religious shift from an impersonal God to one personified, who was portrayed as a healer and transcendent empath,rather than a super-patriarch.

One component of belief in the afterlife likely derived from a mind that tends toward polarities and is programmed by evolution to seek pleasure and avoid pain. However, there is more to the belief in life after death than reward and punishment.

Interestingly, while religious history includes various descriptions of the afterlife, there are, surprisingly, few references to hell. The notion of an afterlife per se was rarely discussed even within the Jewish and Christian faiths. While some sects within Judaism described an ambiguous version of heaven, other sects felt the ultimate reward for faithfulness was to be obtained on earth.

The main reference to heaven and hell in the Old Testament is found in Genesis. In Ezekiel 28:12-17- 28:15-18. Lucifer is expelled from heaven for rebelling against the Lord. It wasn't merely the act of rebellion, however, that sealed his fate. It was the fact that Lucifer had everything going for him, including attractiveness, resources, access to God and a lofty place on the "mountain top." In other words, he was sent down to the pit because there was no excuse for his actions.

The nature of the pit is discussed only in general terms and so is the duration of Lucifer's sentence. Some biblical scholars interpret that chapter in Genesis to mean Lucifer's punishment was temporary. While there was clarity regarding the concept of eternal life in heaven, (particularly by the time Jesus arrived on the scene) there was less clarity regarding what was meant by the phrase "eternal suffering in hell."

Various writers provided commentary on this topic. Bart Ehrman addressed this from a Christian perspective, in his book, *Heaven and Hell: A History of the Afterlife,* He wrote that the concept of life after death likely originated in the Mesopotamian cultures - probably in the Epic of Gilgamesh. In this story the hero contemplates lying in dust and being eaten by worms for all eternity as punishment for his transgressions.

The Greeks were among the first to believe the soul existed apart from the body and would rise into paradise after death. During the period in which Abraham searched for his God, adherents to Hinduism expressed their belief in the complex, winding journey of the soul.

Hindus had a unique take on the afterlife. This was described in the principle of Samara, which dealt with the cycle of life and death. According to this belief, the deceased would be reincarnated in another body. The re-manifestation of their lives would depend on the karma from a previous life. The virtuous would return to benevolent circumstances and assume the form of able bodies. Those whose karma was less impressive would be less favored. There was no heaven in this system but there was a reward and punishment aspect. Thus, the notion of an afterlife cropped up in every civilization, probably well before the beginning of the Bronze Age.

The views of ancient Hebrews differed from that. While keeping constant the idea of a posthumous reward, they concluded that it need not be a heaven in the sky. They saw the afterlife in terms of a return to earth by God for purposes of separating the virtuous from the sinful. He would raise the dead, provide the faithful with earthly rewards, and basically allow the unfaithful to die off. This concept of hell was more an endless void than eternal punishment.

Jews came to believe there was a time and place where God would render judgment, but this did not involve an assignment to paradise or Hades. Nor would it take place in a vacuum. Jews believed that sin - whether based on disobedience, a loss in battle or pagan worship, would result in disasters such as earthquakes, floods, pestilence, and famine. In any case, God's infliction of punishment would take place on earth.

In other words, heaven and hell were effectively considered to be in the Middle East – site being less important than circumstances. Thus, they might have decided there was no need to conjure up celestial sources of pain and pleasure. In that context, Jesus' emphasis on the reward of eternal life in paradise was a revision of an old belief, and highly transformative.

Erhman's discussion on Jesus and the afterlife is fascinating. In reviewing Jesus' sermons, he concluded that his notion of reward and punishment provided a different take on salvation.

As discussed earlier, Jesus was among the first humanists. He believed that merely observing faith through sacrifice and prayer would not get one into paradise. One would also have to engage in acts of compassion and embrace the existence of others, even if they had been enemies in the past. It was an encompassing view of holiness that seemed to add to the tenets of the first commandment. This is indicated in Matthew 12:1.

In this chapter the hungry disciples were reaping grain in the fields. However, it was on the Sabbath, a day of rest, and labor was forbidden. In order to eat the disciples had to act contrary to the law. Upon been criticized for this, Jesus' response was to defend his disciples. In effect, he insisted the Sabbath was made for man, not the other way around. While he referenced the Old Testament in making this assertion to the high priests (Samuel 21:1-6) he was, at that moment, reshaping the Jewish faith into something more humanitarian. At that point the relationship between man and God became one of greater reciprocity.

The Pharisees were livid. Wasn't this blasphemy? Wasn't this upstart Nazarene getting too big for his boots? That question was often asked during his ministry and after his death. It is difficult to say, because Jesus' commentary could be obtuse. He often employed a method reminiscent of the Platonics. He asked the listener to reach his own conclusions then enjoined him in a dialectic discussion.

Was this arrogance - an attempt to outwit those whom he considered intellectually inferior? Nothing in his demeanor suggested that. With respect to his style of communication, there are several possibilities. One possibility is that he was, as some biblical scholars have suggested, a member of the Essene community. The Essenes had introduced a purer, more introspective element into Judaism. Another possibility is that Jesus was enacting his own modified Davidian solution. David fulfilled his mission primarily through military victory, but only because he could match the strength of the enemy.

Jesus was not going to defeat Rome no matter how many tribes he rounded up. Rome had subjugated Greece, Egypt, Gaul, Britannia, present day Spain, all of Italy, Macedonia, Carthage, and various other sovereignties. They were not about to suffer defeat at the hands of poorly organized and meagerly weaponized Jews.

However, Jesus might have realized some victories come only with time. By securing the commitment of as many as possible into his novel theology he might have felt he could convert, rather than defeat not only Romans but the people of other nations as well.

While some historians have viewed his ministry as a failure, Romans were in fact, converting to Christianity in increasing numbers. At the time of his death over fifteen percent of Romans had adopted Christian beliefs, which suggests Emperor Constantine's Edict of Milan in 337 A.D. that eventually centralized the Catholic Church in Rome was neither impulsive nor without precedent. In that context, Jesus' battle plan can be said to have been successful on a massive scale.

Still, another possible explanation is that, as stated in Mark 1:15, Jesus believed the end was coming, that God would destroy all opposed to him while the faithful and compassionate would enter the kingdom of heaven. However, as Ehrman wrote, Jesus did not necessarily believe the kingdom was in the sky. While Isaiah and David had taken their place next to God, that did not mean every person who abided by doctrine would ascend into God's domain.

Part of Erhman's conclusion can be called into question because Jesus was in the habit of promising life after death to the faithful. Whether he meant by that a place in the pantheon inhabited by the holiest of the holy, or some sort of permanent earthly setting in the aftermath of judgment day is hard to tell but the promise of some sort of kingdom was an enticement constantly referenced in his sermons.

The history of thought regarding heaven and hell prior to, during, and after Jesus has been fraught with contradiction, paradox, and irrationality. That might be because accounts of heaven and hell were ambiguously described in virtually all religious texts.

While Buddhism acknowledges the existence of a transcendent place like heaven, the real Buddhist paradise is arguably found within the living soul. In fact, the doctrine espouses that the heavenly endpoint (Dharma) can be reached through meditation. Meanwhile, Hinduism essentially obviates the need for posthumous transport to paradise since the body will eventually be reincarnated and reappear on earth.

The most extensive discussion and the greatest level of disagreement on heaven and hell seems to have occurred among Christian theologians. Some took it more literally than others. For example, St. Augustine of Hippo believed in a concrete, dogmatic version of heaven and hell. He felt God could not forgive the condemned, that once they were dammed sinners would be punished eternally in a lake of fire.

This notion was consistent with an edict issued at the Fifth General Council of the Christian Church in 553 A.D. when it was decided that hell did entail eternal suffering with no possibility of reprieve. St. Augustine's view was so resolute that he believed it applied even to the sudden death of an infant not yet baptized. The idea was that due to original sin, each child is born into the world a sinner. Despite having no conception of right and wrong the child could justifiably be condemned by God.

Others, like St. Anselm argued for a proportional view of punishment. He recognized that sins carry different weight, for example regarding the distinction between venial and mortal sin. Consequently, he felt permanent condemnation to hell was neither applicable nor morally justifiable. His view was very similar to the common law precept in which the punishment should fit the crime.

In an excellent article published in the Stanford Encyclopedia of Philosophy, various interpretations and counter arguments were presented regarding the nature of heaven. A summary within that text suggested the consensus opinion among Christian theologians was that heaven is best described as access to God, wherever HE existed, rather than it being in a specific celestial location.

The diversification of opinions within Christianity over time has possibly been due to the drift from pure theology to the inclusion of philosophy and even science during the Middle Ages. In effect, as religion headed in a rational direction so did notions of the afterlife.

St. Thomas Aquinas felt one could understand the nature of God and gain access to paradise through reason. He lived in a time when the empirical philosophy was gaining momentum in Europe. The capacity to provide greater certainty and provide solutions with more immediacy than faith, made empiricism appealing even to some clerics. The latter might have been put off by Copernicus' heliocentric theory, but many used and even invented mechanical gadgets themselves, reaping considerable profits.

In reviewing many very interesting articles and books on the subject it seems ambiguity regarding the nature of heaven and hell is also because these conceptions are driven by human cognition and emotion, and that we have created, modified, and disagreed about what happens after death because nature provided us with neural hardware that renders such contemplation inevitable.

It is interesting to note that the Stanford article suggested the human conception of hell runs parallel to human experience. Descriptions of hell throughout history make the point. Humans have punished sinners/criminals with fire. Criminals have typically been ostracized, isolated and deprived. Some even received life sentences, which is reflected in the notion of eternal damnation.

Beyond that, however, is the question of where these ideas originate. Indeed, what is it about human nature that requires retribution for offensive acts, particularly those that defy the socio-moral consensus? I believe there are several primary reasons for ideas describing heaven, hell, and the afterlife.

It was mentioned earlier that Neanderthal buried their dead and might have conducted some sort of ceremony as a send-off. Why did that occur? A possible answer can be found in the experimental research on primates.

An illustration of what creates the most intense emotional reaction in great apes (whose brains are complex enough to compare experiences and develop familiarity with the behaviors and presence of group members) comes out of research in a study by Biro and Hunkle in 2010. In that study, chimpanzee mothers continued to carry and attend to their infants after death as if to prolong the life of the deceased. Moreover, in research studies, young chimpanzees exhibited extreme emotional reactions to two stimuli: snakes, and lifeless bodies of other chimps.

The snake reaction is predictable since venomous snakes have preyed on young primates long enough for a social if not genetically driven fear to take hold. But what about the dead body? The prevailing theory is that because the primate brain is large enough to foster neural confusion, it can reach the point where attempts to assimilate confusing stimuli can become highly stressful. When conflict between expectations based on prior experience and actual circumstances becomes intense, cortical circuits exhibit overwhelming hyper-arousal. Because they cannot resolve the conflict a neural downshift occurs, in which lower (midbrain) emotion-dominated circuits take over. If cognitive faculties do not resolve the problem, the emotion programis activated. At that point confusion can turn to terror.

These midbrain circuits first evolved with reptiles and early mammals and are responsible for fight/flight, rage, fear, appetitive and withdrawal behaviors. They do not provide deliberation and other than through emergency responses do not engage in conflict resolution.

With respect to the chimp's reaction to a dead body; having typically seen his fellow in an animate state, the motionless fellow who looks exactly as before, but now exhibits no behavior or energy - would create extreme conflict. Unless the experience occurred so often that the mismatch between life and death could be assimilated, the emotional reaction to a shift from life to lifelessness will be intense.

But while the fear will be intense, the experience won't last. Because the chimp lacks grammatical language and extensive memory, he will typically recover faster than a human. Having the gift/burden of language makes things more complicated. For humans, the experience of another's death and the subsequent period of mourning lasts longer. More enduring solutions will be needed to restore emotional equanimity

. Proportion is involved in this experience. The more one encounters a living person, the greater the internal conflict when that person dies, even more so if there is a strong emotional attachment to that person and/or if that person exhibited energetic, unforgettable characteristics.

Neanderthal's brain was complex enough to experience extreme internal conflict. Whether Neanderthal had language is uncertain, but he had sufficient long-term memory to require a long-term resolution to the death experience. The only way to do that was to keep the person alive - to create a bridge between life and death. Once a creature develops enough brain mass to post-process experiences through memory and language, it becomes more difficult to move on from that experience. Due to the neuronal feedback mechanisms that come with an analytically designed brain, he will tend to suffer after events have come and gone. In other words, the "after" in after-life is fundamentally owing to the evolution of the primate and human brains. At the risk of over dramatizing, it does appear having superior memory enhances both learning and suffering.

Might it be humans are too cerebral to avoid suffering? One could make a case for that. Field researchers have observed primates for decades. Among the most notable were Jane Goodall and Diane Fosse. They have observed surprisingly empathic and intelligent behavior patterns among these species but have not observed the perceptual distortions, breakdowns and incapacity to function seen in humans following severe conflict. While the tendency to ponder and suffer in death's aftermath is reason enough for man to have created the idea of an afterlife, other factors seem involved as well.

The human brain is designed to control both real and imaginary events. It accomplishes that through an internal encoding mechanism that prolongs experience through a process Russian neuro-psychologist Alexander Luria referred to as self-regulatory language. Unlike other species, our reactions and moods are a function of both actual events and our interpretation of events. This model of cognitive dualism was proposed by Roman philosopher Epictetus, and while this belief has its doubters, it does appear an internal replay of experience can determine human reactions as much as the experience itself.

This means man can talk himself into any number of conclusions and emotional reactions. Indeed, the fact that the brain seeks control makes that virtually unavoidable. The human brain operates as essentially a noise buster. Navigating among billions of neuronal connections requires a motivational drive for closure. In evolution that was needed to address the challenge of adapting to various environments during human migrations.

Development of inhibitory networks in the frontal cortex allowed for a neural pause function. That enabled the brain to slow down, deliberate and problem solve, both in the moment and into the future. One component of that includes what psychologist Gordon Allport called functional autonomy, which holds that learning begets learning. Humans are epistemological creatures. We not only can imagine things. We must. We not only invent, reassemble, and reintegrate old and new ideas, we must. That is why human existence has been characterized by an endless pursuit of progress and bold exploration.

Being a curious primate, Homo sapiens is neurologically obligated to explain what happens after death, both because finality is agonizing, and because it arouses our curiosity enough to require conception of an after-life. Thus, fear and creativity seem linked, as per the functions of the human brain, creating a doubly motivating drive to explain, resolve and control as many elements of our internal and external environment as possible.

As for the conceptions of heaven and hell, they seem to reflect the human capacity for proportionate cognition. It is the same neural faculty that fosters mathematical ability and enables us to measure, compare and value experiences. The notion that sinners should be punished and the faithful should be rewarded is simply a moral/verbal version of cognition used in mathematics. It is a socio-moral equation of sorts.

The views of both St. Augustinian and St. Anselm are both reflected in this cognitive template. Augustine's dogmatic, all or none view of punishment, whereby God tolerates no moral gray area is a theosophical version of addition or subtraction. Meanwhile, St. Anselm's proportion paradigm derives from the same cognitive function as the calculation of fractions, decimals, and percentages.

Once again, a prime reason for those tendencies lies in our capacity to extend memory. If one can hold more than one stimulus in mind, those stimuli can be reworked, manipulated, compared, and reshaped in imagination. Another reason is that language-related sequential thought patterns enable us to categorize experiences in terms of "before and after."

The human brain has two cortical hemispheres. The left hemisphere functions grammatically and sequentially. Grammatical communication requires temporality, that is, a sense of before and after as seen in verb conjugations. The temporal structure behind language might be the foundation for what Noam Chomsky and Steven Pinker have referred to as a language instinct. As an instinct, it would tend to foster belief in life after death.

Being bound by temporal cognition leads mankind to extend life beyond its biological parameters, and to include within that posthumous matrix many experiences from real life such as reward, punishment, guilt, right and wrong, suffering and pleasure.

That is not a statement in support of agnosticism, because one could just as easily argue that such beliefs, while influenced by the mechanics of mind, could not be conjured up unless nature endowed the human mind with neuropsychological penchant for believing in an afterlife.

Taking that argument further, one could ask whether Homo sapiens is wired to believe in religious concepts and whether religious beliefs could be considered adaptive cognitive-behavior traits that sustained human (and perhaps hominid) motivation during times of duress, fostered group solidarity through altruism, and deftly converted the primate hierarchy into a more benevolent social format featuring equality and belief in an entity above mankind.

A case could be made that such thought patterns and their behavioral derivatives would have improved our chances of survival over time, that human nature might include a bio-religious component. Just how this could have been manifested over time is worth discussing.

Thoughts and behavior can be adaptive. That is clearly indicated in Homo sapiens' struggles during the Ice Age, when the harsh climate would have had enormous impact on survival.

A crocodile is a cold-blooded creature that cannot function in cold weather. As a reptile, it must raise its body temperature to become active. That requires adequate exposure to the sun long to restore a sufficient metabolic rate. Humans, of course are warm blooded, but our metabolism also changes with the seasons. Homeothermy is a bit of a misnomer. Clinical depression occurs most often in winter and early spring. Activity levels, including hormonal volume increase in warm seasons and decline in cold seasons.

There were times when early human groups dependent on exploration, curiosity and problem solving had to summon high levels of activity even under cold, stifling conditions, for example, during the Upper Pleistocene Ice Age. Some sort of motivational stimulus was needed to overcome metabolic depletion. Is it conceivable that over time, with enhanced language and cognitive enhancement, our species developed the capacity for self-motivation? Is it possible the capacity to "believe," in the face of disheartening experiences enabled our ancestors to prevail and adapt?

There is no specific neuro-physiological structure or gene providing hope. There are, of course, neural connections and reactions providing that potential, but summoning them requires a cognitive impetus. Belief provides that impetus. Any such belief would have to include a sense of present or future control - either by one's own hand or through the intervention of a higher power. Without a capacity to believe in future success, and/or in an entity that, if properly worshiped, would come to the rescue, it would have been difficult to persist, problem-solve and survive. In that context, the argument can be made that there is no true separation between God and mind. Both are psychobiological. Just as one cannot disregard the mind, neither can we disregard the notion of a transcendent figure like God.

CHAPTER 12:
THE TEN COMMANDMENTS

THE BODY OF LAWS BROUGHT FORTH by Moses provided a standard by which human behavior could be judged. Virtually all laws in democratic systems either derive from or are directly in line with those laws. There were, of course, two systems of law in Jewish religion during Moses' time. The Ten Commandments were summary aspects of faith. The second body of laws, the Torah, consisted of the first five Chapters in the Old Testament. The latter provided a more comprehensive, Hammurabi-like set of regulations pertaining to virtually every aspect of life. Some have presumed the Torah is derived from the Sumerian code but in fact it has far more volume and is much more indigenous to one group of people.

The Decalogue (The Ten Commandments) is familiar to everyone within Judeo-Christianity. There have been interpretive variations in the order of those laws over time For example, there are slight differences between the Jewish numbering version in Deuteronomy 5:16-21 and Exodus 20:2-17. Making things even more complex is that there is also a Catholic numbering version. While the differences are not significant, implications in the wording of each version raises questions about intent and the ramifications of each law for Jews and Christians.

The usual list consists of the following

I am the Lord Thy God. Thou shalt have no other gods before me
Thou shalt not make idols
Thou shalt not take the name of the Lord in vain
Remember the Sabbath day to keep it holy
Honor Thy father and Thy mother
Thou shalt not murder
Thou shalt not commit adultery
Thou shalt not steal
Thou shalt not bear false witness against thy neighbor
Thou shalt not covet thy neighbor's house or thy neighbor's wife.

At face value, this body of laws seems like an attempt to maintain order among a specific group of people. The laws are religious, legal, biological, and cultural – a fabulous combination that makes the laws very powerful. Each law offers a set of rules that on one hand, derives from faith and on the other could be understood in terms of the biologically driven survival instinct. The first several laws establish authority, which is perfectly logical because of how human groups function.

We are a species organized by rank, in the form of social hierarchies, and do not differ very much from other primates in that regard. Whether we like it or not, whether we consider any given leader worthy, we seem to gravitate toward A social value system that favors obedience to authority. Indeed, even when humans rebel against an authority it isn't long before we create another social hierarchy. Whether it is based on loyalty, fear of punishment or a symbolic, heartfelt belief in a doctrine that emphasizes the need for authority, the human animal will seek leaders. While social rankings occasionally lead to cruelty and dominance it seems to work well in most instances, especially if the leader's actions prove meritorious.

At the forefront of the social hierarchy are family leaders - the father and mother. If one is inclined to honor and obey them, a tendency to comply and respect authority will prevail. Conversely, rebelling against the parents is a behavior pattern that will often extend to larger society and its mores, including articles of faith.

The relationship between a leader and the general populace can vary. A leader's influence will often depend on his accessibility to individuals within the social order, regardless of rank (because there are always more poor people than rich people), and on whether sub-hierarchies germinate among fractional members of the group. For that reason, any set of laws, and model of leadership must filter through the ranks to be effective.

This has always been the case in legal and religious systems. It is reflected in a cornerstone feature of common law from which many western justice systems derive - the duty of due care. Unless people agree to abstain from harming one another there cannot be a cohesive society. In that context, the commandments prohibiting murder, theft, coveting and bearing false witness are both articles of faith and regulations necessary to sustain social cohesion and in the process, enhance human survival.

The laws in Deuteronomy expand on this bio-religious theme. In the first commandment there is a quid pro quo issued by God, to wit:

I brought you out of the land of Egypt, out of the house of bondage (therefore) you shall have no other gods but me. In this commandment God is basically proposing a version of the social contract that became so integral in modern democracies. It implies that one good deed deserves another. Not that the Lord would have tolerated dissent. It's just that his message seems based more on fairness and reciprocity than dominance per se.

The admonition regarding the worship of other gods raises other issues. It seemed the Lord was wary of the peoples' weakness for symbolic persuasion, and the possibility that art or some other inspirational source could sway people from their loyalty to Him. The urgency conveyed in those words was not necessarily due to God's mistrust of His chosen people. It might have been God's way of acknowledging the times in which these laws were given.

Semites during Moses' time were a diverse group requiring strong persuasion to unify. Some had Egyptian values. Others were exposed to Canaanite deities, perhaps even had ancestry linked to tribes outside the Hebrew line. God could not simply assume they would join in despite Moses' influence.

In short, it was a time and a place calling for strict discipline.

In the Jewish Numbering version there is a similar reward system based on loyalty. Here the Lord explains that honoring one's mother and father will result in one's days being lengthened upon the land of the Lord. It seems clear that the Lord, through Moses, was not just imparting laws, but was also interested in nation-building. While He created the earth, the sky, the water and Adam and Eve in earlier times, this initiative on Mt. Sinai marked a new beginning. It was arguably His second greatest creation.

In that context, it is possible to interpret the Ten Commandments in terms of the need to establish a new society so intact and intractable that it could resist the temptations of other customs, laws, and deities. It was an interesting statement and precursor to a familiar theme issued at various times by Plato, Jesus, and Abraham Lincoln, that... a House divided cannot stand, a house united cannot be moved. It was a time for intensive group solidarity. Thus, the event on Mt. Sinai can be considered both religious and political.

The need for such a momentous religious-historical gesture perhaps reflected the confusion of the times. For thousands of years Homo sapiens lived in close-knit nomadic enclaves. In that context, loyalty was practically a given. The groups didn't need a lawgiver or mountain top inspiration to remain intact. Indeed, the con-sanguine nature of the tribes likely created enough suspicion and mistrust about strangers that group cohesion was difficult to override by outside influence. Nomadic sapiens didn't have to work hard to sustain the genetically bound tribe, and across the millennia that social tendency probably became so natural as to be instinctive - in other words, it would have taken extremely unusual circumstances to break those bonds.

Yet, with the advent of urban settlements, times changed. The tribe was being uncomfortably diversified and seduced by the perks of agricultural permanence, material rewards and the threat of retaliation for not complying with the new norms. God did not appear very often in human history. When He did appear, it was usually in times of dire need. This was one of those times.

One of the interesting questions about the word and presence of God on Mt. Sinai is how attuned He seemed to the nuances of human history. One tends to think about the Lord as rock solid, constant, unwavering and in His perfection, non-malleable.

Yet interestingly, God did change with time and circumstances. On Mt. Sinai He presented himself as totalitarian, pronouncing Himself a jealous god who would exact retribution on those who broke the commandments. That is perhaps why St. Augustine interpreted the Decalogue so dogmatically. He compared the commandments to a stringed musical instrument, noting that if even one string was broken it would ruin the entire song. He might have seen, in the 10th commandment, advocacy of extreme authoritarianism that forbade even the thought of desiring a neighbor's wife, his servants, his oxen, or any of his property.

According to that standard, man's thoughts and imagination were also subject to scrutiny. It was a stringent criterion arguably unmanageable in terms of human nature. It was a tenet later echoed by Jesus in his sermons, yet one that disallowed normal outlets of fantasy, which in many cases can substitute for transgressions and prevent overt acts of sin.

In that context, as good and perfect as They are, neither God nor Jesus was functioning as a psychiatrist in such instances.

God was not always a dogmatist, and certainly has not been depicted as such in recent times. He could be forgiving and accommodating, for example in deciding against sacrificing Abraham's son, Isaac, in what turned out to be merely a test of faith. Meanwhile, despite prohibiting "imaginary" acts of sin, His only son, Jesus forgave a prostitute, dined with a tax collector, and provided medical relief to the servant of a Roman soldier. Was that flexibility a function of changes in the world? Moreover, were those shifts in emphasis meant to be consonant with human nature, therefore acceptable to worshipers?

All of this begs the question of just how detached God has been (and is) from the natural world, and whether humanity has ever been able to completely divest from a homo-indigenous belief in the integration of God and nature, i.e. pantheism. The fact that God could be jealous and dogmatic in one epoch and forgiving in another requires further discussion.

One could assume that since He created the natural world, that world would include Himself, such that, as Aristotle suggested, He was both created and creator. On the other hand, God must be transcendent to avoid being viewed as just another tree or river. Natural elements like those don't control things. They do contribute to the overall stability of the ecology, but they do not "decide." (There is an entity that functions as a decision-making mechanism but that will be discussed later).

Was God above nature or within nature in the mindset of ancient Hebrews? The second commandment answered that question. It was discussed above that God forbade his followers from worshiping other gods, including iconic representations of those gods through imagery. What did this mean? One possibility (referenced earlier) is that He was insisting on Jews being ultra-loyal to one God. Another reason could be that He did not want it presumed that he could be described physically, because that would put limitations on his powers.

In that context, the second commandment could be considered a warning against depictions of God; for example, whether He wore a beard, was tall, portly, slim or muscular, whether He resided on earth, in the sky or in the ocean. That suggests drawing, sculpting, or depicting in physical form any semblance of God would be blasphemous - an extraordinary maxim considering the thousands of depictions of saints, angels and God by artists, clergy, and worshipers over time.

To attain the supreme status necessary to organize, galvanize and sustain the Jewish people this God had to be incorporeal, perhaps even in a sense, non-spiritual. This God would have to be invisible, unknowable, physically indescribable and amorphous, while still being responsible for the entire universe. He had to be both nothing and everything, a vague wisp of wind, then again, embedded in every atom, chemical, force, living being, planet and galaxy that exists in the universe.

It has always been a concept difficult to grasp, which is probably why some within Moses' group questioned His power and, also why Jews subsequently questioned God's existence in times of duress when no divine intervention occurred. The dual qualities of being vague and dogmatic left plenty of room for theologians to revise the nature of God over time.

Jewish law was, of course, more extensive than the Decalogue. Despite some level of disagreement, the consensus is that there are 613 laws in the Torah. The sheer number of dos and don'ts in that body of laws provided stringent criteria on what constituted a moral, faithful Jew back then. Indeed, the Torah is so voluminous that it is hard to imagine every member of the Jewish faith being capable of memorizing, let alone abiding by its laws.

Many of the laws have little to do with adherence to the Lord's will or holiness per se. Some are reiterations of the Decalogue, for example, the mandate to honor the old and the wise, not to add or take away from the basic commandments. Others seem to lack religious premises, for instance, the law on circumcision, on what foods could be eaten (typically excluding consumption of some meat and dairy products) and mandating that tzitzit (tassles) be sewn on the corners of clothing.

As in the Decalogue, many of the laws pertained to social interaction and ultimately to group preservation. One could not curse another, put another to shame, bear a grudge, allow a simple-minded person to stumble on the road, or stand by when another is injured. However, as in Hammurabi's code, the Torah is divided into categories concerned more with social order than spirituality. It deals with marriage, divorce, and the command to go forth and multiply.

Why propagation would be so important religiously and/or socially is an intriguing topic of discussion because, more than any Hebraic concept it entails very basic tribal/ evolutionary themes.

In pre-urban tribes, population dynamics were crucial to survival. A paradox exists in a tribal society because the tribe needs every member to work, hunt, care for offspring, make clothing and perform many other functions. For purposes of genetic continuity tribal beliefs must espouse and ensure that members are alive for as long as possible. A drastic reduction in numbers could result in extinction of the genetic line. Indeed, small group dilution spread across various tribes over time might explain why all hominids except Sapiens disappeared around 20,000 years ago.

Paleoanthropologists have given various reasons why Homo sapiens outlasted all others. Some suggest it was due to superior tool making skills. Others maintain sapiens enjoyed more binding family cohesion. Still others point to superior communication skills.

While all of those are distinct possibilities, it is also conceivable that sexual avarice was (despite waxing taboo in later times) adaptive.

Archeologists' discoveries suggest some human groups were prone to sculpting female figures with extraordinarily large breasts and the modern obsession with sexuality through various media supports the notion that sexual preoccupation is built into human nature. Ensuring adequate birth rates -especially given what were likely high infant mortality rates, was urgent enough to turn reproduction into an aspect of faith and law.

Nature has a way of correcting population errors, as though operating like a cybernetic machine. Small populations will tend toward hyper-sexuality, while larger populations might experience depleted sexual interest. In that context, while modern, populous human societies espouse sexual restraint, low population, early human societies might have believed in the opposite. In the transition from nomadism to urbanism, groups like the wandering tribes from Akaad, Babylon, and Sumeria (later called Hebrews) might have fortified their natural sexual inclinations by inserting religious beliefs into the mix.

The command...go forth and multiply, was not peculiar to Hebrews. It is, as discussed above, a natural mechanism designed to keep genetic groupings intact. For example, a study in *Heredity*, showed that a population decline typically leads to increased hermaphroditism in some types of animals (whereby organisms can be both male and female). Such a transformation would increase the likelihood of reproduction.

That natural/sexual correction factor might explain why humans have the largest sexual organs of any primate. Having greater sperm production would tend to increase the birth rate and considering the tribal wanderlust of Homo sapiens over the millennia, this might have become a stop gap against extinction. Did this ancient motive trickle down to urban Homo – and to the Torah?

If so, it was perhaps dually conceived, first through an unconscious process in accord with the notion of survival motives of the "selfish gene." Second, as manifest through language and cognition, as conceived in religious texts such as the Old Testament.

Still, another example of how population dynamics produce changes in behavior are the r and k selected parenting styles of various organisms. Caretaking and mating practices seem to operate by principles similar to economic inflation. The history of economics has shown that the value of money and declines in proportion to its availability. Similarly, when the population of any organism is diminished, its parenting style will tend to be more attentive and fastidious. This hyper-attentive style is called k selected parenting. Conversely, r-selected parenting occurs with a high yield birth rate. In that instance, the value of each offspring declines, which prompts a relaxed parenting style. The higher numbers in effect lessen the value of any given offspring as a kind of bio-inflation.

Human beings are not relegated to instinct so our appraisals of the value of life can result from cognitive/rational assessments and cultural influence as well as biological drives. At the same time, since we are natural beings, it seems unlikely that we are completely exempt from biological influence.

Humans can bypass instinct. We have politics and a superego to modulate primal urges. However, we cannot completely override them. Freud wrote that we channel instincts to accommodate the demands of society but cannot eliminate them because we need the energy derived from the id to do so.

As ironic as it sounds, the primal urge is a governing force of our actions and feelings, the clay from which human probity, creativity and social responsibility are sculpted. That would seem to have influenced the religious and legal systems over time. Indeed, it is possible all legal and religious systems, including the Torah, the codes of Solon and Justinian, the Magna Carta and the Declaration of Independence featured human minds asking God, or some transcendent figure for guidance. That might explain why there are so many common principles embedded in legal and religious documents.

The laws in the Torah regarding marriage, social behavior and property are reminiscent of Hammurabi's legal system. Whether they were derived from the latter is moot, simply because any set of laws designed to create order in any given society would have to include a process by which man and God cooperated to improve the viability of a worshipful species.

Another connection between the Torah (and the Jewish state of mind) and the psychobiology of man is seen in the list of rituals involving cleanliness. The laws could have been designed for religious purposes and, also to reinforce tribal distinction - a bit like a secret code that only members of an exclusive club can know. It was surely not based on snobbishness, for the Torah warned against debasing, shaming, or lying about strangers. In fact, it is an amazingly eclectic document requiring Jews to show compassion toward Gentiles as well as their own people.

Despite that, tribal solidarity is emphasized in certain parts of the Torah; for example, regarding propagation within the group and regarding their own laws and customs taking preference above religions and customs of other peoples.

There are contradictions within the Torah, which, on one hand, compels respect for others, and on the other hand, requires keeping strangers at bay to preserve cohesion, prosperity, and the genetic/cultural exclusivity of Jews. However, it is predominantly altruistic. In many ways, this wonderful set of laws seems to have captured not only the essence of Jewish morality but the nature of the times as well.

It was a time of transition. Tribal integrity was compromised by a drift toward diversity. It created irresolution between the tribe and the nation, between genetics and politics, between monotheism and an ecclesiastic view of the world. It seems man had to required to accommodate two central but incompatible socio-biological urges: the need to survive during times in which large armies had the advantage in war, and the need to preserve the genetic purity of the group, which required the exercise of idiosyncratic rituals, language, and actions.

That raises questions about human nature. If we are prone to preserving specific genetic lines does that mean we are predisposed to bias, discriminatory practices, and even racist sentiments? Moreover, is it inherently wrong to be prejudiced, or are we so guided (vexed?) by human nature that overcoming such sentiments requires a victory of nurture over nature?

In any case, creation of the Torah marked the beginning of a new era. It was perhaps the first time that politics and faith battled for the hearts and minds of a creature with certain instincts who was nonetheless able, through neo-cortical brain functions to override atavistic impulses. As Freud suggested, human history might well be described as an attempt to resolve a conflict between natural man (the id) and learned man (the ego).

It is conceivable, that through His laws and social perspective God was an integral part of this process. Perhaps he spent the ages trying to teach His subjects how to solve that bio-political puzzle. If so, He would have to be viewed as a duality: both the creator of human nature and orchestrator of man's imaginative capacity to override human nature.

In any case, it seems to this writer that the idea of God must be something woven into both mind and nature. The question is, what force / deity could be both cognitive and spiritual, both omni-influential and non-material?

For many centuries the true nature of God was pondered, discussed, and altered. At times, a consensus was reached, then subsequently challenged. It remains a fascinating and open question.

CHAPTER 13:
THE DUALITY

THE ANCIENT GREEKS' IMPRINT on modern religion, science and politics is, in many ways, difficult to describe. Just why a relatively small, nation state came to influence so many cultures is a complex question. Being a seafaring nation certainly provided them an opportunity to spread their ideas. Also, their relatively democratic outlook fostered creativity resulting from open discussion. Still, within that society there were as many barriers as opportunities. For example, city states had to compete for dominance within narrow geographic boundaries. Spartans clashed with Athenians and Corinthians clashed with both. War often precludes creativity.

Unification of several hundred civic entities (forming the Delian League) did occur after the victory over the Persians at Plataea in 478 B.C. Yet conflict seems to have run parallel to invention and philosophy within the craggy borders of these fledgling societies. Such turmoil has typically led to one of two possibilities in the course of history. Either chaos results from exacerbation of tribal hostilities or, as with David in Israel, a central leader espouses a core doctrine acceptable to disparate tribes. The latter did not happen in Greece, which raises the question of why they attained such cultural prominence.

One historical influence came from Egypt, but while its education system, described in Plato's Principles of Early Education utilized dialectic methods, the general culture in

Egypt could not have prompted the many ideas that emanated from Greek minds, especially given the scope of those ideas and the many Greek schools of thought.

Among the early, pre-Socratic philosophers, were members of the Silesia school whose most illustrious exponent, Thales, believed everything in nature arose from a single source - water. He is considered by some the first scientist, because he adopted the practice of making predictions based on measurement. His method was employed in Egypt for purposes of measuring water levels on the Nile prior to planning irrigation projects.

Taking a cue from Thales, Xenophanes came to believe all elements in the universe had a natural rather than divine cause. He did believe in God, but only in a holistic context. He felt God was woven into nature - as if some sort of geophysical algorithm and source of all divine decisions and interventions. In essence he was really presenting an idiosyncratic version of pantheism.

Many have written about the origins and influences on Greek philosophy, which covered areas including mathematics, politics, biology, physics, ethics, ontology, aesthetics, and astronomy, but there seems not to have been any true antecedent to Greek brilliance. Historian Martin Litchfield West wrote that the Greeks taught themselves to reason, which seems the most reasonable explanation for the intellectual big bang spreading forth from the Aegean peninsula.

The term 'Greek epistemology' seems an accurate way of capturing the essence of their collective accomplishments, since knowledge seeking was their ultimate quest. The Greek zeitgeist was replete with ideas about virtually everything in the cosmos. Pythagoras (discoverer of the famous geometric formula) had more to offer the world than just one theorem. He also attempted to merge religion with reason; a task that would filter down through the ages and influence, among others, St. Thomas Aquinas.

Pythagoras' quest was quintessentially human. Since Homo sapiens first developed a forebrain capable of collating, comparing, and contrasting sense impressions, there had been two essential competing ideas. One held that nature derived from transcendent forces. The other held that nature was so expansive yet self-contained in its wondrous majesty that it merely appeared to be transcendent. Pythagoras was one of the Greek philosophers whosought resolution to this duality.

Another was Parmenides, who argued that nature is a completely unified entity. Because of that, he felt movement is impossible, that motion and space cannot exist because they are woven together, thus could only alter their position relative to one another – resulting in no net movement. Although highly abstract, this idea was prescient, in view of Albert Einstein's discovery of special relativity many centuries later. Still, this was only one of many provocative ideas circulating around the various Greek city states.

A fascinating debate also took place between the Atomists and Pluralists, who were arguably the early architects of the empirical movement. The former believed nature could be understood in terms of a single or limited number of forces - a mindset similar to the current belief in a theory of everything. Some said the single entity was water, others, fire, while still others believed it was something in the atmosphere called "ether." The Pluralists believed what we saw was all there is - each item or phenomenon having its own functional and structural essence.

In reviewing the history of pre-scientific and post - scientific thought, it often seems that humans are inclined to gravitate toward relatively fixed cognitive templates; and are, in a sense unable to really explore the natural world objectively. Indeed, it is possible that because the human brain adapted to a broad but circumscribed set of environments, we might not be able to extend beyond that. Just as our hearing is not as acute as a dog's, it is possible human cognition is restricted by evolutionary adaptive parameters.

It seems possible that, for all the new discoveries over time we just might be circling endlessly around central themes embedded within mind that are as much a function of our evolutionary wiring as of the objective aspects of the natural world. If so, that would suggest (perhaps somewhat cynically) that even with the most advanced scientific experimental designs and statistical formulas, we can only discover what is within our cognitive engrams.

Does this, somewhat skeptical opinion coincide with the uncertainty principle in quantum physics, which holds that the observer affects the outcome of the experiment? More discussion on that later.

The ancient Greeks carried things much further. Parmenides and the pre-Socratics were just the beginning. While Pythagoras searched for heaven on earth and God within nature, Socrates took the next step toward unification of God, morality, and nature. Roman orator Cicero said of Socrates: "He was the first to bring philosophy down from the heavens, placed it in cities, introduced it into families, and obliged us to examine life and morals, and good and evil."

Socrates was the ultimate iconoclast, with such impactful ideas that he managed to alienate even the democratically inclined powers in Athens, leading to his imprisonment and eventual suicide. He was a Jesus-like figure before the real one came along, and he had his Paul of Tarsus in the person of Plato. The latter captured Socrates' life and philosophy in three main works: *The Republic, The Laws,* and *The Statesman.*

In addition to acting as a Socratic spokesman, Plato introduced a philosophical concept bridging the ancient and modern world, called metaphysics. He and his illustrious pupil Aristotle engaged in debate on this subject, particularly as it pertained to the pet projects of each - the quest for the perfect version of all things. The reasoning behind this theory of "evolution into perfection" was based on the idea of teleology - a futuristic developmental process

The argument revolved around the problem of transitions. Plato felt all things in nature existed on two levels. One was concrete. The other was a spiritual and perfect version of itself. This included people. In contrast, Aristotle believed the perfect version of any object or being could be found within that entity, that one did not have to look for a transcendent version of a horse, rock, or person to understand its essence. One only had to strip away the imperfections within objects to find purity - and God

Aristotle's version required an empirical view of nature (the notion that one could understand things through observation). Yet, in adopting a teleological perspective he anticipated many aspects of both future scientific inquiry and religious contemplation.

The fact that Aristotelian philosophy seemed to dovetail with Darwin's notion of survival of the fittest, the Buddhist concept of moral perfection through attainment of Dharma, the Judeo-Christian notion of redemption through confession, and the Hindu belief in reincarnation through karma is perhaps one reason his ideas were revered for thousands of years. It was why Avicenna called him "The Master" and why St. Thomas Aquinas referred to him singularly as "The Philosopher."

Even more amazing is that while this marvelous thinking was going on, the Greeks, including Plato and Aristotle still worshiped a plethora of pagan gods; each of whom had human-like flaws as well as divine powers. Each had stories of origin, marriage, family lines and endpoints to their existence. It appears, somewhat paradoxically, that the Greeks managed to describe religion in earthly, human terms while still describing the gods in a spiritual manner.

Through Greek philosophy the relation between man and God was transformed. Such a transformation would end up sounding the trumpet for everlasting change in human society.

Gods would remain important, but a duality was in play. God and man were potentially on more equal footing than at any time in human history. Religion was still a central factor in ancient Greece, but gods were so decentralized and, in many ways humanized, that it precluded the presence of an absolute deistic authority. That opened the door for human cognition to explain the essence of nature. It seems the Greeks were fearless when it came to partaking of the tree of knowledge, and as a result, created a more egalitarian relationship among man, God and nature.

They still required an ultimate authority, simply because, for all their discoveries and theories, they could not find the answers to ultimate questions. Consequently, many of the philosophical, medical, and mathematical innovators attributed their ideas and discoveries to divine guidance. In fact, that held true all the way up to Isaac Newton's invention of calculus and his mathematical formula for gravitational attraction. The problem was that with so many gods overseeing various aspects of life - and often burdened by their own trials and tribulations, divine attribution became problematic for the Greeks.

In many ways, the Greek pantheon resembles that of other religious systems that emphasize a line of descent. Whether this was purely religious and cultural, or the result of the bio-natural tendency among humans to favor their own genetic line is hard to tell. However, since the religious, legal, and political concepts we espouse derive from our biological dispositions, both elements were probably involved.

The Greeks had their theosophical hierarchy, with stories accompanying the rise of each. It was a different take on divine origin. In the Old Testament, God was considered a constant. He had no beginning or end and there was no story behind His emergence. In fact, He was not said to have "emerged." Instead, the entire universe did, at His behest.

The Greeks proposed a developmental version of God. For example, Zeus was the main god, but to attain that status he had to battle against the Giants and the Titans. He and his comrades, including his son, Hercules, eventually emerged victorious. As a result, Zeus became the prime deity who governed the heavens, created laws for his human subjects, and unlike distant gods in other religions who intervened only occasionally, Zeus was said to interact with, protect and nurture individuals and families and protect strangers traveling to uncertain destinations.

As an alpha deity, Zeus was said to have a healthy libido. He had a lovely wife, Hera, but to her consternation, he frequently spread his seed with other goddesses and human females. One of his brothers was Poseidon, ruler of the kingdom of the sea. Another was Hades, the god of the underworld (he was also referred to as "Pluto.") He was a dark figure, far less popular than other gods because he represented the specter of death. He was also a somewhat cynical deity because in order to pass into his underworld one had to pay a toll - lest his guard dog, Cerberus launch an attack.

Zeus' family line extended well beyond that. Another of his sons was Apollo, the god of light, music, poetry, healing and prophecy. Then there was Apollo's sister, Artemis, the goddess of hunting and the wilderness who was also said to protect mothers from disease during childbirth. Interestingly, Hera was deemed both Zeus' wife and sister. This incestuous relationship might have derived from the Egyptian practice of inter-marriage to maintain the purity of the blood line. Regardless of prior influence, it seems the Greek pantheon of gods was consistent with so many aspects of human nature (within nature) that it could be deemed, like most other systems, pantheistic.

While the gods had various responsibilities and personalities their provisions to mankind were no different from what is provided in the natural world, for example normal childbirth, securing the seas (the Geek economy was heavily reliant on seafaring trade and the fishing industry), control of the heavens, from which came sun, rain and warmth to contribute to crop yield, general health and the light needed to read, write and visualize in an era without electricity.

The gods also provided intrigue, including sexual scandals, conflict among themselves and commoners and greed. Teleology notwithstanding, none of the gods was perfect. They charged entry fees for the underworld, consorted with the wives of lay persons, and Ares, the God of war was at one point, charged with war crimes due to his bellicose actions.

The question was raised previously as to how the Greeks, as originators of logic, empiricism, and explanations of how the cosmos worked, could adhere to such a Byzantine system of worship.

Human cognitive was undoubtedly responsible for their brilliance and religious ambiguity. While personifying the gods brought them closer to the people - perhaps an unavoidable byproduct of a democratic mindset, the Greeks were inexorably drawn to the same quest that occupied the minds of Jews, Zoroastrians. Buddhists, Christians, Muslims, and Hindus- the search for an answer to how and why the world works as it does.

It seems possible that all belief systems over time emanated from that search. Finding the ultimate truth, the reason for existence, resolving the question of why life consists of progress and setbacks, and wondering who or what has regulated, parented, protected, and occasionally punished Homo sapiens through time.

We have also sought definitive answers to the question of right and wrong. Since the human approach to problem solving has often relied on an (animist) assumption that nature can be understood through para-social empathy; for example, assuming trees have feelings, that the wind has a temperament, that the motives of animals are the same as ours, it makes sense to seek knowledge by internalizing the world around us. Interestingly, while that might seem a dubious methodology, the fact is humans are part of nature. While our brains are larger than all others relative to body mass, much of our behavior can only be interpreted in natural terms. We don't use present-sense cognition to "decide" to be jealous, to "plan" to be competitive, to "self-monitor" in order to stabilize the body's homeostatic functions. In many ways, it might turn out that we can only learn from the inside out - even the most rigorous experimental design being merely another manifestation of the imagination.

CHAPTER 14:
THE FERTILE CRESCENT

DESPITE DIFFERENCES IN BELIEFS and cultures with each successive political/religious movement, it is conceivable that all of us have been participants in a long relay race; each culture handing the baton to the next in the quest to find ultimate answers. Some adherents seem to have carried a heavier load and run a longer, more grueling leg through history.

The Hebrews of Moses certainly did, in creating an ingenious merger between law and religion. Perhaps, through faith and intuition they realized that for man to become civilized would require the influence of both. It worked well. Bless the Jews for their moral innovation, blending empathy, internal morality, and legal exactitude.

Buddhists and Christians then came along, creating a shift from faith based on obedience and fear of retribution, to faith based on the humanistic idea that it is intrinsically rewarding and moral to care beyond the core group (and in the case of Buddhism, for all creatures on earth). The positivism arising from Jesus and Buddha was inspirational, but they were leaders espousing a creed that was also germinating in other minds.

Prior to that, the Greeks suggested that since both people and gods could be fallible, what really counted was the notion of merit. According to their teleological belief system, one didn't have to be perfect. One merely had to work toward improvement; perfection being a long-term ideal. The implication was that one did not have to engage in pomp and circumstance or hypnotic, whirling dervish rituals to get in touch with honor, ethics, or virtue, because when all is said and done, these ends could be achieved through reason and introspection.

Through Greek influence, morality turned inward. Through the efforts of Aristotle and the rationalists, there was a vehicle by which to reach that goal - not just through prayer or religious sacrifice, but through use of a magnificent tool as important to human progress through the ages as the wheel or the printing press - the syllogism.

Before science was modernized in the laboratory, it was carried out within the neural conglomerates of the human brain. All it took was for mankind to delve deeply into that realm, to consider himself worthy enough to seek answers, not from God but from within, and engage in the risky business of figuring things out on his own.

Obviously that internal drift didn't provide all the answers, but it was a formidable leg in the historical relay race. Since history is, in part, a cognitive continuum, each movement deriving from a mix of prior ideas, the historical sequence is not always predictable, notwithstanding Hegel's triadic theory and Marx and Engels' concept of dialectic materialism.

Odd things happen, and interestingly, the next cultural movement offered a dual contribution. One element was intently emotional, spiritual, and parochial. The other was non-spiritual, detached and objective, and became the most accurate way for humans to understand the true nature of the cosmos. The source of both was the same: a group of people referring to themselves as Arabs. One of their creations was the Islamic faith, the other, the mathematical decimal system.

Islam begins with the story of Muhammad. It seems appropriate that he followed the paths of Abraham, Moses, David and Jesus. His journey from the cave, after the Archangel

Gabriel provided him with his monotheistic inspiration to his death in 632 C.E. contained trials and tribulations, similar to those of his predecessors.

Muhammad experienced the same monotheistic conversion as Abraham. He united twelve tribes, as did David. In espousing a new version of faith, he invited his neighbors' scorn, as had Jesus. Yet there was more to him than that.

Muhammad also served as political organizer, businessman, father of many and central prophet of a faith so similar in doctrine to the Christian and Jewish faiths that it is hard to imagine why the conflict among Jews, Muslims and Christians has persisted right up tomodern times.

Like Jesus, much of Muhammad's childhood has been lost in historical accounts. However, it appears his early years were a bit less pleasant than those of Jesus. The latter certainly experienced hard times. His parents had to flee to Egypt to avoid Herod's wrath and Jesus was born into impoverished circumstances. However, his parents were there for him and lived to see him emerge as an adult. Muhammad's father died before he was born, and he was orphaned at the age of six. During his early years, much of the Arabian Peninsula was populated by Meccans and Medians. The area was mostly arid, save a few concentrated areas with arable soil. The scarcity of farmable land necessitated a nomadic existence. That made unification of various peoples to a single belief system difficult. Despite the ongoing transition from nomadism to urbanism, tribalism was the predominant social structure in many areas of what is now called the Middle East. Muhammad reacted to tribal diversity in ways similar, to the reactions of Abraham, which made sense, since he believed he was descended from Abraham's son, Ishmael.

Pivoting off that legacy, he sought to convey the religious beliefs, not only of Abraham but also Moses and Jesus. He also acknowledged a tie-in to the original man - Adam. However, the times had changed. While Abraham traveled in search of God,

Moses departed from a dominant empire to build a new religious society, and Jesus tried to overcome Roman oppression through the promise of a higher kingdom, Muhammad was more prepared for and inclined toward conquest from the outset. Like Zeus, he battled his way to preeminence, first by uniting the tribes in Medina, then with victories against the Meccan tribes. As he gathered more influence and power, he began to spread the word and create a new culture, as though a combination of Paul of Tarsus and Alexander the Great. Soon the beliefs, architecture and general culture of the Muslim faith were disseminated all over the Middle East.

Despite his military prowess, Muhammad seems to have been primarily inclined toward peace. He forged several treaties, requiring that tribes abstain from battle for years at a time. That enabled him to travel freely and safely so he could impart the words of the God he called Allah. His sermons/instructions to people derived directly from consultations with his God, which took place in caves as well as in his mind and heart. For the most part, the message was humanistic and appears to have been multiply influenced.

At the time, Jewish, Christian and Buddhist theological principles were familiar to most, despite a climate of tribal detachment. Muhammad seemed to accept ideas similar to the ideas of Moses, Buddha and Jesus, particularly regarding the shift from obedience to compassion.

The words of Allah came forth. The weak were to be protected. Wives (while subject to husbandry control) were to be treated kindly. It was important to help and give to the poor There was a bit less reference to love - as compared to the words of Jesus, but that was because Muhammad's doctrine was more about deeds than contemplation. The emphasis in Islam was in the doing, not the feeling - at least not predominantly.

Such an action-based faith was destined to reap dividends. Muslims were like Buddhists with respect to their emotional and compassionate orientation, but quite different in that internal mental states were not considered sufficient by Muslims. One suspects that while Allah's instructions to his prophet were obeyed, like Jesus, David, Moses and Buddha, Muhammad probably inserted his own personality into the process.

The doctrine was adapted over time - although the demands of Muslims on how to act toward their fellow humans has remained essentially the same. The interesting aspect is that due to Alexander's conquests, Greek philosophy was fully available to all societies in the Middle East by the 7th century of the Common Era.

In his campaigns, Alexander's goal was not to simply subjugate territories. He also sought to advance the cultures of the conquered, likely because he was tutored by Aristotle and believed in teleology; the idea that everything in nature contained potential perfection that could be brought forth through an idealistic mindset. He brought along scholars, scribes, engineers, and artists as well as military personnel, including the brilliant general Ptolemy. The Egyptian city of Alexandria was a central point from which new ideas were spreading. Its residents included philosophers, mathematicians, physicians and artists and it contained the largest library in the world.

While the Islamic belief system resembled that of the Jewish and Christian faiths, it tended toward a more authoritarian structure. Islam has only one true prophet, though, as in Christianity, there are lower-level prophets, and higher level "rasuls." However, it has no organizational hierarchy like the Catholic Church. As a result, it became a faith at once pristine in its message and somewhat vagarious in terms of its leadership. Still, the message Muhammad imparted was pure and very much within the humanist model that had been building in human society for centuries.

It seems, due to his role in establishing the Islamic faith, Muhammad had to be all things to all people. He was the writer of treaties and conveyor of legal principles, the father of many, due to his having a multitude of wives, a politician who could persuade, a soldier who could conquer, a diplomat who could live comfortably with Christians and Jews when he and his followers needed a place to hide during various battles among hostile Arab tribes. He had to be pragmatic as well as spiritual in order to galvanize his people and solidify their faith. That was one significant contribution from the Arab world.

There was another. Tracing the history of mathematical development is difficult because it developed in bits and pieces and in various places over time. In fact, the origin of math reasoning goes back over four million years to the origin of arboreal primates. Math is really a symbolic version of spatial reasoning. To think in terms of concepts like "more," "less," or a progressive sequence of numbers, requires some understanding of spatial relations. There is no better spatial analyst than the arboreal primate, who must calculate distance between branches, the height and arc of its leap and the number of barriers from one branch to another. It must also estimate the length and firmness of tree limbs in relation to all those factors. And while the worst that can happen to a math-challenged human is to flunk a test, the primate's calculations have life and death ramifications.

Math isn't just a set of formulas written on a blackboard with umpteen symbols in an M.I.T. classroom. It is also the center fielder in baseball judging where a fly ball will land, how quickly he should time his first step, and when to finally reach out with his glove to make the catch. It is ubiquitous in the existence of every primate, and certainly of Homo sapiens. One could ask why, given such evolutionary importance math did not come to fruition until the 8th century. To answer that question requires discussion of math usage in the earliest civilizations.

In Egypt, measurement became crucial for two reasons; first to determine flooding trends in the Nile for purposes of planning irrigation projects, second to guide the construction of monstrous pyramids and other buildings which they put up prolifically. One way in which they cultivated these skills was through an advanced education system enabling young children to have strong math interests and capabilities. Plato visited Egypt on various occasions and in his book, *Laws,* Plato described the method of teaching.He wrote:

"They teach children at the same age that they are learning to read and write. The teaching takes the form of pleasant games like dividing out apples and flowers, now in large groups, then in small groups. Then they take vessels of gold, silver and brass and sort them out. They freely adapt the games to the numbers available. In this way, they enable the students to learn about the movements of armies and supplies. They learn how to manage a household. The pupils are more alert and in touch with reality. They learn how to measure and count. In this way they are better able to deal with things around them."

Some historical accounts suggest Pythagoras taught the Egyptians how to take measurements in constructing the pyramids, but from Plato's observations it seems they had a firm grip on the essentials of mathematics from the outset.

The problem with early math systems was twofold. For one thing, in most cultures, math symbols were embedded within the language system. This was true with Greeks, Romans and to an extent, Egyptians. One exception was the Sumerian system, which enabled them to calculate the distances of heavenly bodies with a fair amount of accuracy. Yet initially math was concrete. For the most part, it could not be used to develop formulas for spatial and temporal relationships or statistical models by which to determine how measurements change over time. That stalled invention, advanced civic engineering, and other requirements of progress. In effect, math had not yet been completely merged with the imagination. That's where the Arabs entered the picture. Their system is now used by virtually everyone on earth and the mechanics behind its development are fascinating.

Prior to engaging in discussion, it is important to consider the history of mathematics and its relation to God before the decimal system came about because like other fields of inquiry, it was tied to religion for a longtime.

In The Origin of Species, Charles Darwin wrote about a process called a conversion. This referred to the fact that an evolutionary trait that initially enhanced survival could, in changing circumstances, be used for other purposes and turn out to be adaptive. That seems to have occurred in the case of Homo sapiens in the transition from nomadic to urban life.

Living in permanent agricultural settlements did more than provide political and social change. With more down time, cognitive energy previously used for the urgent planning of migrations, hunts and territorial adaptation hit a wall. Sedentary life might be more stifling, but it also facilitates creativity. The potentially imaginative nomadic human brain was free to "roam" in urban settings. The human animal might have reduced his travel habits but his brain, having evolved as an omni-adaptive organ was still on the move; not geographically, but internally.

While there is little evidence to suggest the human brain evolved further during the agricultural movement, its wiring likely did change toward a greater frontal influence on other brain sites. That would have made for a more introspective, imaginative, internalized, and worrisome mind. At that point, rumination could have substituted for attention previously devoted to the stress and strain of survival in various environments. The human brain was then able to divert its neural hardware into "could be," "should be", "might be" appraisals. From that neural restructuring came a plethora of new ideas.

This was a watershed occurrence. Beyond freeing up the imagination, the dense population in urban settings led to competition among ideas, greater opportunity to compare notes and a tendency for small groups to collaborate. Since the Sumerian culture was the first large urban settlement, it was appropriate that it all started there. Written language was invented, featuring a wedge-shaped letter system known as cuneiform. This system was less sophisticated than the Egyptian system of hieroglyphics because the latter was phonetic as well as visual. But it was an impressive start. Along with that came the invention of the wheel, the plow, irrigation methods and other innovations. Also, in a large population it was necessary to develop systems of apportionment for resources, census taking andother administrative functions.

Thus, basic mathematics seems to have had its inception somewhere around the middle fourth millennium B.C. The earliest features were cumbersome. In the Sumerian system a small clay cone represented the number 1. A larger version represented the number 60.

Over time, the Sumerian math system improved to a point where it could be used to measure long distance phenomena. Since crop yield depended on weather conditions and the phases of the moon influenced tides and celestial predictive elements, they developed methods for measuring celestial events. This might have been the source of their worship of the moon (Sim) as a prime deity.

Their system had a base of 60 - which meant one had to stretch out each calculation quite a way before being able to group number sequences into sets. It had no decimal point by which to create distinctions among fractional quantities. It was laborious, but far more effective than that of the Egyptians and even the Romans and Greeks in later times.

One of the most significant features in this system was a concept that later came to be known as zero. It did not initially create much impact except as an abstract starting point. In fact, the ancients did not initially know what to do with this idea.

Meanwhile, math plodded along. Counting, representing quantities, doling out resources, taking the census - all these functions were served well by the existing math systems leading up to the Common Era. The one thing lacking was a system of abstract mathematics. Without that type of math, calculations and projections were necessarily tied to concrete events. This was useful. but not theoretical enough to lead Homo sapiens to the next phase of discovery.

Interestingly, the mathematical systems conceived at that point seemed in parallel with religious practices. No matter how hard these people tried they could not quite think outside the bounds of religious and natural influence. The gods influenced the hunt, the seasons, and the sea but these were all phenomena within the purview of nature, which meant in terms of human cognition, God really did not occupy a space that was separate from the rest of the cosmos. He was inexorably embedded in that process, not a singular, distinct orchestrator of nature, but as the ancient pantheists, Greek Atomists and Buddhists believed, an entity operating from within the confines of nature. More specifically, God was not viewed as a first cause, but as both cause and effect, part of a larger whole and all the whole, and logically, necessarily, amorphous.

Centuries went by and math remained stagnant. However, as always, there was a causal historical sequence of events. For a long time, there wasn't enough mixing of cultures (the true engine of human invention) for mathematical breakthroughs to emerge. It was the Romans, as nation builders, catalysts of cultural advancement and occasional orchestrators of destruction who came to the plate with bat in hand.

In the first millennium Islam was making significant cultural, religious, and military inroads in the Middle East and beyond. They had their culture, which, aside from Islam itself, was derived from that of Egypt, Babylon, and other nearby civilizations. A new entity originated at the time. It was called Arabia and its culture was nomadic, tribal, and insular. Initially, there was a lack of influence from outside the area, which led to a period of relative stagnation. Then in 529 B.C.E. Roman emperor Justinian, fearful of sophisticated and potentially revolutionary ideas of Greek philosophers, destroyed both

Plato's Academy and Aristotle's Lyceum. That was followed by a period of censorship. As a result, many brilliant Greek thinkers left town and headed east. They continued to teach, and as their ideas gradually seeped into the eastern mindset new combinations of ideas came into being. It set the stage for a massive intellectual explosion.

The Greeks brought abstract views on truth, ethics, proportionate thinking, empiricism, cosmic contemplation to the concrete mathematical concepts of the eastern peoples. A new hybrid emerged from the confluence.

. The concept of logic was juxtaposed on the idea of the number zero. Zeno's abstract notion of sequential immovability, Pythagoras' theorem, Euclid's new version of geometry - all these ideas merged into something new.

This did not happen at once, or fluidly. Arabs and Greeks did not initially trust one another. In 642 Arabs destroyed the library at Alexandria. However mutual respect eventually seeped into the intellectual environment - at least among the philosophers and artists (which seems to be typical for our species, even during periods of armed conflict).

While new ideas were developing, the perceived origin of each remained rooted in core religious beliefs. The number zero was seen as emanating from God's grace and creation. Euclid attributed his geometric concepts to God. Aristotle viewed math as consisting of two separate elements. One was the concrete correlation of a number...for example three apples. Aristotle referred to this object-symbol tie-in as the heap and believed it derived from man. The other, the monad, was the abstract version of a number; for example, just *three* without an object tie-in. Since the monad had no explicit connection to a specific object it was presumed by Aristotle to have divine essence.

Most Greek philosophers and mathematicians felt similarly. Euclid was convinced math was so spiritual as to be incomprehensible to the human mind. As a result, he never bothered to offer proofs of his theorems. Because he was not the source of theory, he probably felt any criticism directed at his formulas should logically be directed at the Lord - a testimony to his political deftness.

Math was still within the grasp of theology up to the 8th century until an Arab named Muhamet ibn al Khwarizmi combined ideas and came up with the most economic and functional system ever devised. He was not the sole forerunner of modern mathematics, but he was a uniquely gifted one (he also invented algebra). He had company.

The Chinese had developed a system with a base of nine and an ingenious gadget that precluded having to use one's finger to calculate. They called it the abacus. The brilliant Omar Kayam made significant contributions as well, and across the ocean in a land few knew existed, the Mayans had also stumbled upon the concept of zero. Yet, while it seems the entire human race was racing toward mathematical resolution it was Khwarizmi who, by adopting elements of Hindu mathematics, wrote the first book describing the decimal system. It was the first breakthrough.

The significance of the Arabic numeral system is twofold. First, it simplified the process of grouping, so increasingly larger sets of numbers could be calculated without overwhelming the minds of the calculators. Second, it created a clearer separation between nature and God that would eventually give rise to the idea that God and science are different sources by which to determine cause and effect.

As mentioned earlier, it seems ironic that the source of deistic-mathematical detachment was also the source of a fervent, unwavering religious belief system known as Islam. There is no historical evidence to suggest Khwarizmi, Kayam or any other Muslim mathematician viewed their foray into in modern mathematics as thebeginning of an atheistic mindset.

The fact that there wasn't an attempt by Muslims to address the separation between math and God resulting from the decimal system is interesting because in some sense the 8th century Middle East might have represented a new version of Eden. In the following epochs many would be tempted by the tree knowledge and faith would be tested repeatedly. Eventually a schism between religion and science did occur, not to the complete exclusion of faith, but certainly headed in that direction.

The decimal system was facile, and it eliminated the mystery seen in mathematics. It's use of zero was not really that complex. Zero was not conceived in the same way that negative numbers are today, and surely not to the same degree of abstraction as modern algebra. It was just a reference point - like the period in a sentence. One could count to ten (quite naturally derived from the human hand) then stop and put a zero aside the last number. The zero signaled the end of the set, which made it easy to tabulate the number of sets. For example, 1,2.3.4.5.6.7.8.9, followed by a 0 (indicating the number 10) signified that 1 set (encoded by the zero) had been completed. Then...11,12,13,14,15,16,17,18,19, and 2(0) signifying that two sets have been completed. Such a system facilitated the use of addition and subtraction as well as multiplication and division.

The system allowed one to work with large sets of numbers because it was memory friendly. Ten fingers on the hand were no longer a required mathematical frame of reference. Nor was there a need for the abacus, which counted to nine, then left it to memory to keep track of the sets and no symbol to refer to.

A whole new world was being created on both microcosmic and macrocosmic levels. Since it seemed possible to measure the dimensions of rivers, mountains, deserts, celestial bodies and conceivably the universe itself, it might have dawned on a few that God was no longer a necessary guide to knowledge and prediction.

The world did not wax agnostic just because of the Arabic numeral system. and of course, it hasn't yet, even though many modern scientists discard stories of creation in favor of explanations from the field of physics. Perhaps, that is because, as Karen Armstrong suggested, it is native to our species to believe in a God construct. It seems clear that evolution handed us a social hierarchical mandate forcing us to look beyond ourselves for supervision, protection and understanding. It might also be that the human brain is not fully capable of understanding how the cosmos works, not just in terms of gravitational attractions, forces, and quantum phenomena, but also in terms of how and why we behave in certain ways, why our moods can vary widely, why love and hate are so ergonomically intertwined and why one can so easily shift toward the other.

We have not yet attained understanding of the totality of nature and might never do so. Yet, due to the closure-seeking propensities in the voluminous human brain we seem obliged to search for explanations. Precise mathematics did not answer those questions but there was a sea change in human thought after Khwarizmi's text was published, and a bi-modal quest for the ultimate religious/scientific theory soon took up a head of steam. It led to the empirical philosophy.

The first wave of intellectual revolutionaries was ambivalent. Like the first rock and rollers: Elvis, Little Richard, and Jerry Lee Lewis, they were torn between their enthusiasm for the new trend and the guilt they felt about its irreligious implications. Some were inventors, iconoclasts, and antagonists, but not atheists. They struggled to find new relationships among natural phenomena to determine why objects fall from certain heights at certain rates of speed, why the motion of planets seemed curiously at odds with the presumed centrality of the earth in the cosmos.

With each discovery there was an attempt to reconcile man-made discovery with God's preeminence. The new breed must have felt uneasy at times, because while they were faithful to their God, by the act of discovery, they appeared to commit the sin of pride - the same act that led to Adam and Eve's demotion and Lucifer's expulsion in Genesis. In a way, these explorers were not so much riding into modernity on a white stallion as tip toeing gently into an epoch portending man's tenuous ascendancy in the natural world.

CHAPTER 15:
GALILEO AND NEWTON

HUMAN PROGRESS HAS NEVER BEEN CONTINUOUS. Like evolution, it is punctuated, with leaps and bounds followed by periods of stagnation. In some instances, the causes of alternating growth and regression have been clear; particularly when natural forces are involved. In an abstract, but not completely subjective context, it might appear nature frowns upon continuous human progress, perhaps, as occurred in the Garden of Eden, recognizing that the tree of knowledge can be a deceptive source of gratification, by tricking man into orchestrating his own demise through excess.

If it weren't for a devastating set of circumstances, mankind's progress might have been more advanced than now. Toward the end of the Dark Ages, things were changing technologically, politically, scientifically, and artistically. Yet two regressive circumstances served as checks and balances on human advancement. One was the decline of the Roman Empire and its previously galvanizing influence on Euro-African culture. The second was an outbreak of various plagues in Asia and Europe.

Rome had done what all great civilizations do; made itself necessary for over a thousand years. Its language, politics, culture, art, religion, and civic engineering principles create a blueprint for many nations. Building bridges and roads, organizing the military, constructing temples and marketplaces, enacting laws, naming cities, even creating religious hierarchies, made it more than a highly influential society. It was tantamount to a paternal entity, governing, disciplining, and bestowing pride through association with "The Eternal City." Yet, ultimately, its power and influence waned.

In 476 Odoacer put the final nail in the coffin of Rome's western empire. It wasn't a single battle that did the trick. Rome had undergone entropy over time. Corruption by a series of insane leaders, a massive immigration of Goths who fled from the Huns into what they considered safe-haven, declining economics, and a depleted military unable to control its vast territory led to decline of the western empire. While the eastern empire would go on for a while, the Roman imprint was fading fast.

Because Rome was a strong influence, its dilution led to a splintering in western society. As with the decline of the U.S.S.R. in the 20th century, Rome's decline led to revolutions among smaller entities interested in seizing power and establishing their own governments. None was particularly good at it, so the model of the empire whittled down to local government models. Towns like Paris, London and Venice became relatively independent.

Loyalty and power were ceded to local rulers. Fiefs cropped up, each with their own economies and militia. Just as Rome's ambitious, culture-centric philosophy made the world a smaller place, its decline made the world much larger and diverse. Moreover, the Latin language evolved into various dialects and eventually into completely new languages. Rome's policy of delegating governing authority to local rulers in its diverse territories led to the emergence of local power. Since, in a small sovereignty, it is neither practical nor necessary to devise a republican format - as did Rome for much of its existence, the power of leaders in local towns reached a point where the seeds of fascism in Europe were being planted. It seems local rule often became totalitarian due to lack of effective check and balance mechanisms.

Less regulation and vigilance by a central governing body, coupled with economic decline, made travel in areas between small sovereign entities easier.

Because local governments tended to be oppressive it also became necessary to leave the cities.

Such circumstances turned out to be both good and bad. Increased travel led to a broader exchange of ideas, making it possible for the entire western world to become more learned. However, since travel often involves greater transmission of diseases there was also an exchange of microbes that nearly took down all of Europe and Asia.

Plagues had always accompanied travel. and were periodic. There were six major plagues between the 6th and 18th centuries. The first occurred during the reign of Emperor Justinian in 541 A.D. In reporting on this event, the historian Procopius noted that on average 10,000 people a day died during this epidemic. It first broke out in Constantinople, then spread rapidly, keeping pace with global travel and social contact.

Still, the people of Eurasia carried on. There would be subsequent epidemics leading to a loss of many lives. Among them were the Italian plague of 1629 which lasted two years, the Plague of Marseilles in 1720, and a pandemic in 1855 that originated in the Yunnan province of China. Each of these pandemics wrought devastation but none had as much impact as the Black (Bubonic) Plague of 1347.

It appears Italian sailors carried this back from Crimea, and it was, to that point perhaps the most catastrophic event ever recorded in human history. The Bubonic plague lasted five years and killed 50 million people. However, its impact went beyond misery and death. Physicians tried frantically to understand and treat this worldwide malady through techniques available at the time.

Being versed in the science of the time, most were familiar with the works of Aristotle, Ptolemy, Pythagoras and certainly Galen. They were using math concepts to augment their diagnostic methods and prior to 1347 likely viewed the future as bright. It might have even seemed to them that dependence on God was about to come to an end, that Homo sapiens was about to fly out of the nest on the wings of the empirical movement. They might have surmised that in time, God would have become a luxury, a spiritual accoutrement employed primarily to spice up celebratory occasions.

Then came the year 1347. After it became clear that bloodletting with leeches, lance boiling and bathing in vinegar would not work, physicians resorted to a twofold approach. First, they stopped treating the sick to avoid becoming infected. Second, they got down on their knees and prayed, convinced that something this horrible could only have arisen from the wrath of God.

In effect, the attempt to replace faith with science failed. It seems epistemological Homo sapiens had to concede defeat to an omniscient deity for the time being. That swing of events changed the world as much as the decimal system or the printing press. Humility before the Lord was once again in vogue. Uncertainty was embraced, and while scientists searched for biological causes of the plague, they did not have much luck. Whatever entity was leaping from mosquitoes to rats to people was either too small to be seen or too ephemeral to be considered part of the biological world. It appeared to be as invisible and insidious as Satan himself.

Since the will of God (not merely his existence) was once again a paramount consideration, Homo sapiens had to go back to the drawing board. Fearing the post-Eden quest for knowledge might be in some sense blasphemous, man was forced to bifurcate his attitude toward discovery by apportioning part of the process to himself, his method and his math, and part to the Lord.

Since Aristotle was considered a prime source of wisdom during the Medieval period his mathematical distinction between the heap and the monad seemed a useful means of exploring the natural world without offending God's sensibilities. But another of his ideas that helped Medieval people learn to accommodate God and science was a notion he referred to as The Unmoved Mover.

The concept had enormous future implications. It conjured up the image of a formless, eternal entity that did not engage actively in the natural process. This concept did not involve a parting of the seas or sculpting people out of clay as with Moses's God, or bestowing higher status to mere humans through sexual relations with commoners, as with Zeus.

Despite implied limitations, Aristotle's prime entity was somehow able to make everything happen. This concept is seldom referenced in either religious or science books today but in principle it created a potential explanation of ultimate truth for both clergy and scientists.

The idea of an Unmoved Mover simply meant there was an uber-regulatory force, not in itself caused by anything, yet causal in the way all things work. It was arguably a first foray into the search for a Unified Field Theory. For theologians, it provided a way to prove the existence of God. For empiricists, it was a way to support determinism and continued scientific exploration. As a result of its breadth, the idea of an Unmoved Mover created a potential merger between faith and science and helped Homo sapiens find his epistemological golden mean.

It also had a psychological impact by providing two options. Mankind could now be submissive and humble in the knowledge that there was something out there making the decisions for all the world yet be bold enough to explore the essence and parameters of this governing force. He might never find it - God's will might remain beyond man's comprehension for all time. But he could certainly try without remorse and did; leading right up to Heisenberg's principle of uncertainty, which suggests man is so entrenched in the natural world that he can't possibly be an objective observer.

It seems the world needed a philosopher to cope with the plagues, overcome the arrogance presumed to emanate from human curiosity, and at the same time acknowledge the necessity of submitting to a hierarchical presence.

Few ideas were available during the Middle Ages to suit those purposes. Thus, while the empirical movement eventually disproved many of Aristotle's theories, those ideas - and particularly the notion that God could be a theo-natural, blended entity and still regulate the universe enabled mankind to embark on a path to Enlightenment.

In order to appreciate Aristotle's brilliance - especially for his time - requires some discussion of his writing on Metaphysics. In Book XI of that treatise, he described the universe in a way quite consonant with modern physics. For him, the primary universal principle was motion. He wrote that the universe operates through causation, that everything moves and changes as a result of an interaction with something else. For example: A would interact with B and cause it to move or change its properties. Then B would interact with C and cause it to move or change. It was a bounce and rebound concept that coincided roughly with phenomena inherent in the modern theory of quantum mechanics, which brings to mind the old saying that there is nothing new under the sun.

Quantum principles came about through the work of Einstein, Planck, Heisenberg, Feynman, Schrödinger, and a host of others. It started off by challenging the idea that fields or waves existed in the universe and were responsible for causation. After Einstein's work on the photoelectric effect, it became clear that discrete quanta, in the form of photons bombarding electrons, produced the energy needed to release electrons.

Interestingly, while the word "quantum" seems mysterious, it simply refers to the fact that a discrete unit affects properties of force and matter, including the motion and energy level of other discrete units, and that the cosmos is primarily made up of collisions and absorptions between and among those discrete entities.

Aristotle's theory went beyond that. He assumed that if one traced movement back in time, there would be a point where the first cause had to be motionless. He could not have predicted that special relativity theory would provide a way for even the first link in the movement chain to be itself in motion. However, while special relativity refuted Aristotle's notion of a first cause, the latter was on to something. For example, Einstein also discovered that there is an entity in the universe that is unchanging, regardless of what is happening around it - the speed of light. Thus, the general idea that action (and all that it creates) cannot occur without a constant universal anchor point has remained in play.

While speculative, if not downright fanciful, it might be tempting to consider light a facsimile of the Unmoved Mover. After all, it is necessary in the creation of life. It provides energy for plants and animals, without which cellular metabolism could not occur. It heats the earth enough for life forms to exist. That gives the earth preeminence in some sense. It is also responsible for producing many of the biochemicals that create proteins and acids that build tissues and provide mitochondrial reactions allowing us to move, emote and think. In that context, it is not surprising that the sun was worshiped in many ancient religions.

Even more significant, in terms of the ostensible co-evolution of mind and religious faith was Aristotle's methodology. He did not travel out to the desert, seek solace in caves or dwell on the shores of narrow rivers to baptize a flock of worshipers. He relied on the syllogism, which was among the first of many attempts to combine reason with faith. By itself, that would have been enough to establish Aristotle's lofty status in the domains of faith, science and virtually every other field. Yet, he offered even more.

His next logical endpoint lay in his belief that motion was everlasting, that once something was moved it would continue moving ad infinitum. Once again, he was both wrong and right about that. Motion is potentially everlasting according to the law of inertia. Interference by another force or object is required to stop the motion. Thus, in its purest state, movement is eternal. He further reasoned that since movement was caused by something it had to be moved by some other object or force and therefore the original source of the movement could be determined by tracing its movements back in time. But since backward movement could not go on forever there had to be a point where the cause was motionless – a point of ultimate stability.

While this sounds like cold physics, Aristotle inserted love into the picture. By attributing this process to God, he allowed a divine entity to be worshiped, not just by believers but by all people concerned with the nature of the universe. He felt through their orbits in space the planets and stars were imitating the Unmoved Mover as a reflection of its influence - much like gravity creates pan-influence among various celestial bodies.

No one prior to that, came close to formulating a scio-theological synthesis, and arguably, no one has since. In modern physics there have been references to the "God particle" (the Higgs boson - a force particle that is assumed to imbue particles with mass). Also, Einstein argued against the idea of a quantum - probabilistic (rather than lawful) model of the universe by stating, "God does not play dice with the universe." However, Einstein also expressed skepticism about the existence of a creator. Thus, it was Aristotle who paved the way for future exploration into the integrative relationship between God and nature.

An interesting point - which has been woven through this book, is that it seems no matter how spiritual, empirical, mathematical, or subjective our beliefs, we might be inexorably drawn to a pantheistic model. Somehow, attempts to reach beyond the natural world for explanations have always fallen short.

While Aristotle did not create an entirely new, integrative scio-religious concept he did espouse a viewpoint by which inquisitive man and omnipotent God might reach agreement on the details of their respective roles. It was an important nexus.

Renaissance artists, musicians, entrepreneurs, and philosophers derived much of their inspiration from religion. Michelangelo's most notable works were his sculpture of David and his paintings in the Sistine Chapel. Much of the music, including lyrics and melody were tributes to Christ. The Roman Empire became the Holy Roman Empire. Meanwhile, Gutenberg's invention of the printing press, which could have disseminated ideas on many topics, was used primarily to print copies of the Bible.

Still, while Aristotle provided a link, many people were either too devoted or too fearful to opt for moral and intellectual independence. As Copernicus and others discovered, the empirical philosophy was still considered blasphemous in many quarters. It was a time in which Europe was referred to as Christendom. A time in which politics and religion banded together in a power play that would foment acts of reformation and rebellion and lead to the persecution of "sinners" accused of heresy for ill-defined actions. It was also a time of supreme irony.

Although Aristotle provided an inroad to an integrative theosophy, the Christian faith was leaning in the direction of extremism. The one modulating factor was a non-spiritual item known as the gadget.

Scientific discoveries in physics, chemistry and geometry showed inventors how to make better compasses, clocks, and metallurgical compounds. Because these scientifically derived objects were not spiritual in nature and certainly did not challenge the notion of God, they were accepted, bought, and sold. Some members of the clergy were inventors themselves. A climate of invention made life more convenient and bootstrapped economies throughout Christendom.

And of course, economics was crucial to societal growth. It might seem logical to talk about the Renaissance as a period of creative exploration, but artists needed sponsors. Seafaring trade ventures required funding. Economics has always been a foundation of societal progress.

The difference in attitude toward mechanical (vs. theoretical) science was clear to Galileo Galilei. On one hand he earned fortune and acclaim for his invention of the telescope (which he called the spyglass) because it made travel on the seas more efficient and enabled seafarers to spot, avoid and/or prepare for pirates before an attack. On the other hand, he was persecuted by the Church for his acceptance of heliocentric theory.

Profit kept science in play despite prevailing religious fervor. As a result, the Christian Church was beginning to focus on economics. Indeed, the wealth accumulated by the church enabled the clergy to build universities where they could control the subject matter and teach students in areas deemed integral to faith; for example, music, literature, art. Math was not generally taught due to its potential to supersede God's primacy. Galileo's family was quite aware of this trend.

Vincento Galilei was a skeptic by nature who was fearful that his talented son's abilities would be stifled by the rigid training in Christian schools. Still, the boy was allowed to attend school in his early years at a monastery in Vallombrosa. Following Galileo's graduation, his father made every attempt to divert him from training in the clergy. Galileo did not resist - his father was his idol, and the boy was a chip off the old block. He enrolled in the University of Pisa where his father hoped he would become a physician.

That is where the boy's vision for his future and that of his father conflicted. At the University Galileo turned into something of a jack of all trades - the type who was also a master of several. He got interested in physics and math and ended up teaching astronomy. The combination of Vincento's iconoclastic nature and his own intellectual restlessness set Galileo on a path to discovery. Eventually he would clash with the church, but in his youth, he was about as free in his curiosity as any man alive.

Eventually it became all about the heavens. The core belief among the newly galvanized religious/political establishment was that the earth was at the center of the universe. There was no other formal opinion at the time. Certainly, the Polish astronomer, Copernicus had determined that was not true, but he was reluctant to put it on paper. Eventually he did, in his book Celestial Bodies, but he did not come forth to submit heliocentric theory until virtually on his deathbed.

Galileo might not have set out to break the rules, but he decided nonetheless to let his observations and calculations be his guide. Back then there was little in the way of public education, so new ideas took a while to spread to a point where they could change public opinion. One reason for the discrepancy between the elite and uneducated public was that almost all academic writings were published in Latin.

Galileo had a disciplined mind, and he was stubborn. The quality that made him such a phenomenal scientist also got him into trouble on occasion. It wasn't just with the authorities, as per his initial refusal to publicly refute heliocentric theory. It also showed up in his contentious relationship with his daughter, Virginia Livia, whom he sent to a convent to prevent her sexual awakening, and with some of his students. (He filed a lawsuit against a young man he felt had stolen his idea for the spyglass).

He was less contentious toward his fellow scientists, particularly Tycho Brahe, who inspired Galileo through his invention of an early model of the telescope. Galileo knew both the value and limitations of Brahe's invention, and as an avid astronomer addicted to the notion of truth by observation, he wanted to look further out than anyone previously had. His solution came about from his superior knowledge of optics. He tinkered with combinations of concave and convex lenses and discovered he could magnify objects by a factor of fifty. It led to the cosmic shot heard around the world.

On the night of January 7. 1610 Galileo peered out into the heavens and had a momentous vision. First, he noticed numerous irregularities and signs of explosive events in various planets, particularly Jupiter. It looked as though planetary formations were shaped by a series of events over time rather than being created in one fell swoop. It suggested an evolutionary process, which seemed to challenge the account in Genesis. Being empirical by nature he felt this must also apply to earth. Galileo conducted further observations and made discoveries with profound implications for astronomy, physics, religion and perhaps everything else that mattered to human beings.

He compared the motion of the moons around Jupiter, made calculations, and decided there was a relationship between the mass of a body and the closeness and speed with which smaller objects revolved around it. His lengthy observations of this attractive process convinced him that a force was involved in bringing small objects in contact with larger objects and that this arrangement seemed constant. He called this phenomenon gravity.

He took the next step by assuming this principle pertained to the entire cosmos, that a natural, not divine force was not only governing the structure and function of the universe but could be understood and quantified.

To another, it might have seemed dangerous to submit an idea that diminished God's status. The question was whether his theory of gravity would be accepted in the religious community.

Of course, it wasn't. Galileo had a dear friend who rose in the ranks of the Catholic church to become pope. He took the name Urban VII. Previously, Galileo had written a book, The Assayer, as a tribute to his friend. However, the discovery of gravity turned the friendship upside down. It didn't help matters that Galileo published his famous book, *Dialogues*

Concerning Two Sciences, in Italian, to make his findings (and their implications) available to the public.

Urban had to choose between an old friend and God. It was not a difficult decision. Galileo was imprisoned - albeit in relatively humane conditions. He was not put in chains but placed in what might be called nowadays, house arrest. He was asked to recant. Galileo resisted initially, and only after becoming drained by illness and isolation did he issue a reluctant refutation.

It didn't matter. The cat was out of the bag. Enormous disparities in wealth and status, as well as the cynical alliance between the clergy and politicians created mistrust in the public, particularly among reformers who felt faith had drifted too far in a materialistic, secular direction.

It seems the discovery of gravity was one of the most significant events in human history. On one hand it was just a guy taking note of planetary attractions based on mass and distance On the other hand, it became such an important feature of the natural world that it would dominate the field of science for thousands of years.

Despite all the brilliant minds that came along since Galileo, including Newton, Planck, Heisenberg, Maxwell, Einstein, Fermi, Feynman, Oppenheimer, Hubble, Penrose and Hawking, no one has been able to explain how this force (generally described as being "weak") could not only influence the entire structure of the universe but also appear so detached from other forces to have virtually its own source of governance.

The collision/absorption model of quantum physics - the notion that discrete quanta create force through interactions does not pertain to gravity. The real "weak force," which is responsible for decay through the release of radiation from a nucleus, occurs through interactions between particles. The strong force does so as well. It creates matter by binding together the nuclei of particles. Meanwhile, the electromagnetic force creates energy through particle interactions (attractions and repulsions). Each of these forces is characterized by a quantum push-bounce effect.

But there is no apparent bounce or push with gravity, which has an opposite pull-attraction effect. In that sense, gravity seems not to act as a force at all. Einstein initially thought gravity was a version of electromagnetism but abandoned that idea due to disconfirming evidence.

No one has found the equivalent of an electron, photon or neutron that creates gravitational attraction, (the graviton is still a hypothetical construct). As alluded to above, gravity is described as a weak, negative force, because all other positive forces create an outward thrust rather than an attraction.

For example: picture a stick of dynamite going off. The thrust will be outward. Imagine the unfortunate occurrence of an automobile crashing into a guard rail. The rail will dent outward from the point of impact. Gravity never pushes things out, has no thrust. There seems not to be anything quantum about it. It only draws objects inward in proportion to the relative mass of the objects and distance between them. It is a common phenomenon known to most grade school students, yet it is still a mystery despite being so overtly functional.

Gravity is perhaps the most uncommon of common phenomena. We rely on it to stand, walk, do the dishes, toss a baseball with dad in the backyard, and plan a trip to the moon. As Newton wrote on several occasions, we don't really know why it does what it does. Had Galileo's discovery occurred in previous times "Gravity" might have been a name given to a preeminent Greek or Egyptian god. In studying it, they would have seen its regulatory influence was even more profound than that of the sun and water. After all - seasons, the weight of objects, tides, the construction of bridges and buildings, the cause of illness to human beings burdened by excessive weight, the pain of arthritis, for that matter all the resources necessary for life to exist, are a function of the law of gravity. Among ancients, there would have undoubtedly been ceremonial tributes, prayers and sacrificial rites devoted to this force.

Of course, the ancients did not know of this law, though some Sumerians might have engaged in erudite speculation. While Galileo discovered that gravity exists, he was unable to determine its operations with mathematical precision. It was left to another man to cross that threshold. At the time of his discovery, religion was a bit less antagonistic toward science, particularly in England.

While the field of science typically moves ahead in fits and starts there were two periods, each lasting little more than a year, that revolutionized not only science but the human condition. The second of these has been referred to as "annus mirabilis." It was the year covering 1905-1906, when a young doctoral student and patent worker in Switzerland published papers on special relativity, the photoelectric effect and the relation of mass and energy in the iconic formula: $E = mc^2$. The former was devoted to gravitational phenomena. The latter came out of the ground-breaking discovery that light was not a wave but consisted of particles. That, of course, was Albert Einstein.

The other occurred much earlier and was just as revolutionary. In 1665-66 Isaac Newton came up with the mathematical equation for gravity. It stated that the gravitational force was proportional to the product of two masses and inversely proportional to the distance between them. In other words, more massive objects attract less massive objects, and the force of the attraction depends on both the mass differential and how far one object is from the other. He also came up with a new theory of light and invented calculus in that same narrow time frame.

Calculus was invented because Newton realized science was, up to that point, concerned primarily with the relation of objects and forces in static circumstances. Prior to Newton's invention there was no way to measure the changing acceleration of objects. In studying gravity, Newton came to realize that the speed of objects increased at different rates upon their descent. The calculus accounted for changes in variables and made mathematical predictions on that basis. His system replaced steady state math, and this enabled scientists to predict statistical trends and progressions over time when relationships among trends were not constant. It is still the prime mechanism by which to predict elections, timelines for space travel and other things. In effect, Newton's new math combined estimation with mathematical precision and created a whole new way to describe the universe.

It was a marvelous tool that one might imagine offered the opportunity to detach completely from worship of God. After all, calculus was about ascertaining the future. One could look at a trend, gauge slight changes over time, develop a statistical model and determine with a fair amount of accuracy how events would transpire over time; for example, weather patterns, the projected height of people based on childhood stature, gene patterns, even the likelihood of stock prices rising and falling.

In ancient times, it was left to prophets and soothsayers to predict the future. By their reckoning, an omen today could mean victory in war tomorrow. Yet they had no mechanism by which to determine the likelihood that victory in one battle would be followed by victory in the next battle. It was assumed only God could make that determination.

Newton changed all that. In the aftermath of his discovery, it might have appeared Homo sapiens had the world at his feet, that while God was a useful concept in the early development of the human species He was no longer needed. Early on, Homo sapiens had been a submissive child, requiring guidance from a God/parent, but with a new tool at his disposal he was entering epistemological adulthood. Therefore, some might have felt it unnecessary to hold religious beliefs

The problem was that Newton himself did not believe in detachment from God. Like Euclid before him, he attributed the magic of his discovery to disclosures from the Lord. Isaac

Newton might have done the leg work, but by his reckoning it was God's benevolence that enabled him to develop calculus, the gravity equation, and the idea that white light contained all colors of the spectrum.

Such a mindset not only preserved God's primacy. It also implied a fusion between mind and God; implying that, despite the existence of neuronal capacities giving rise to thought and action, manifest knowledge had to pass through a divine filter to emerge. One could argue that this meant God, said to be everywhere, was most fundamentally situated in the brain. It was a notion that perhaps gave rise to Descartes' belief that the soulwas situated in the brain's pineal gland.

It does not appear Newton's deistic adherence was based on fear. He seems to have accepted that man was not above nature, therefore could not really uncover the majesty of the cosmos on his own. Nature was too complex, too powerful, too vast to be truly understood.

Science was impressive. It changed lives, revealed principles, made predictions, and measured large quantities, but in the final analysis, the scientific explosion during the Renaissance and Enlightenment seemed just another failed attempt at God - displacement.

Empiricism – the belief that truth and fact can only be discovered through observation, had been on the move before with each new discovery and with the rise of scientific and philosophical theorists, but in general the wise, hairless upright walker wasn't quite buying it. The image of God, entrenched in history and planted in the heart and mind, seemed indelible.

CHAPTER 16:
St. THOMAS AQUINAS

THE AGE OF ENLIGHTENMENT WAS A TIME when Homo sapiens didn't quite know what to do with himself. It must have been both frustrating and tantalizing. The questions before mankind were: Shall we choose God or science? Do we need the Lord to guide us into the future, determine right from wrong, let Him decide whether a war will be won or lost, or whether the harvest will yield a good crop? Or do we take a chance on calculating and observing our way to self-sufficiency through the scientific method?

It was frustrating, not just because of discoveries by Newton and Galileo but also because a Frenchman named Rene Descartes came up with a method providing the possibility of prediction and control without veering off into atheistic arrogance.

Rene Descartes is often described as co-founder (along with Galileo) of modern science. His unique contribution was a principle known as universal doubt. It refers to the fact that in research projects the observer must initially assume that his hypothesis is wrong. That forces the researcher to evaluate results in an honest and meticulous manner and avoid leaps of faith and personal bias. In effect, the principle, which was later called the null hypothesis, purported to take anthropocentricity out of the picture.

The principle wouldn't last forever. Quantum physics would eventually place man right back into the deterministic equation through the anthropic principle, but it created a standard of inquiry that served us well and led to solid advancements in the sciences.

While universal doubt allowed for bold exploration, the issue of ultimate control remained unresolved during the 17th century. Science and religion were both firmly entrenched in the culture. The question was whether two ideas could share responsibility for creation and simultaneously be deemed a "first cause."

Over time, neither idea avoided skepticism. God's infrequent intervention has occasionally led to abandonment of faith among worshipers, while science has occasionally pushed us into an uncertain future in which we ended up creating gadgets and devices that destroyed the people who came to depend on them. Indeed, a glance at the history of inventions seems to indicate that all have killed people. Vaccines, automobiles, airplanes, guns, and electrical appliances have all had a destructive component, even if they were mostly beneficial.

Early on, Aristotle provided possible answers to the God/science conundrum. He believed God and nature could be co-worshipped, but he was no longer in the picture and guys like that didn't come around very often. A new voice was needed.

The 13th century featured a re-invigoration of Aristotelian thought because similar ideas were introduced, allowing people torn between science and religion to take solace. It began with a gifted theologian. Like Aristotle, who was versed in several disciplines and did not accept absolute exclusivity between God and nature. This Italian theologian provided a facsimile of Aristotle's Unmoved Mover and issued a comprehensive response to the question of primacy.

His name was St. Thomas Aquinas. He was a brilliant bridge builder and one of the few who could have served as a guide to modern times. He wrote on topics so diverse that subsequent philosophers have had difficulty determining whether his ideas had a secular or divine basis. In fact, he dealt with both, and prior to the time of Galileo, Descartes, and Newton his theosophy proved more inspiring than most of his theological predecessors.

Aquinas arrived on the scene during a time when reason and faith clashed, and his ability to blend the two gave him great stature.

Born in 1225 in Roccasecca, Italy, a monastery town, his family eventually moved to Naples, where he eventually enrolled in the University of Naples. At that time, novel ideas were germinating in the universities and Aquinas encountered rather interesting individuals during his education. By then, Aristotle had become a champion of the church because his ideas were compatible with those of the Catholic faith. Yet there was more to it than compatibility. In an oblique sort of way, the clergy were drawn to Aristotle's ideas on metaphysics and dialectic discussions of doctrine.

It was not surprising. The church built most of the universities, and with lots of young students there was bound to be exploration of new, controversial ideas. Although the church created a curriculum favorable to doctrine, they probably knew debate would eventually break out. St. Thomas was caught up in the search for new ideas and for closure regarding conflicting old ideas. However, his approach was a bit different. Along with his admiration for Aristotle, he had an unwavering devotion to faith. His complex writings and beliefs attest to that. He was not necessarily interested in casting religion strictly in terms of reason and logic, but rather in using logic to validate religious concepts and the existence of God. Instead of empiricism taking over from faith, he believed faith could be proved using rational, empirical methods.

A central question prevailed during his lifetime. Can man discover God and the course of nature on his own or did he have to rely on lessons imparted by God? It was a conflict pitting the active learner against the obedient worshiper - the same one that has vexed our species since we first developed the capacities for language and abstract reasoning.

Aquinas addressed this question by dividing secular and divine subject matter into two parts. In *Summa Theologica*, he discussed the difference between mind and body. His arguments borrowed from Aristotle's writings. He came to believe an object could be material in its essence, and completely derived from nature (for example the human body), yet be composed of substances that, while collectively making up the object, might not in themselves be material. This concept allowed him to deem reason (based on the conclusion that the body was material) and faith (which did not require material essence) equally valid.

That enabled philosophers and scientists to accept the existence of the soul. The idea was that the soul was incorporated into the body (as Descartes also presumed) but was itself devoid of material content. It provided a new integrative standard. He proposed that the soul and other spiritual entities couldn't be seen, touched, weighed, or measured. That made his paradigm acceptable to clergy and the faithful because of its ostensible proof of man's immortality, the miracles and divinely created tenets like original sin, the afterlife, and heaven itself. It was also acceptable to empiricists and secular philosophers because it acknowledged that people died, that problems could be solved through science and math, that the cause-effect operations in nature could be studied and understood objectively.

It was an impressive philosophical position that conveyed a significant influence on how people viewed Judeo-Christianity as well as the emerging empirical philosophy.

The method wasn't entirely new. Aristotle had conceptualized things in similar fashion. However, he did not have the burden of reviving the Catholic faith during a time when iconoclasm was in play.

St. Thomas Aquinas developed theories on so many topics that covering the depth of his postulates in one chapter would be not only impossible but pointless. It would seem more interesting to discuss his contributions in terms of the age-old conflict regarding the issue of control. For example: Who has control, man, or God? Moreover, who should have it, God, man or a hierarchical figure somewhere in between?

All of this speaks to more than Aquinas' or Aristotle's contributions. More generally, it concerns the evolution of human cognition and how humanity chose to navigate through the rough waters of faith and reason.

Every human society has had its theology. Nowadays. In the face of scientific progress, some have refuted the existence of God. Modern day agnostics might claim man no longer needs a deity to prescribe moral principles. After all, we have laws to do that. The obvious answer to skeptics is that the laws and ethical standards by which we live have all been shaped by religious doctrine. All the commandments became part and parcel of legal systems - albeit modified over time. Furthermore, even agnostics today live and behave according to moral tenets derived from religion. It might be convenient to say that morality is ingrained in human nature whether God exists. However, since moral principles derive from religion the two can't be viewed separately.

In effect, we have no way of knowing how Homo sapiens would have turned out if not for the guidance of religion along the way - especially since religious worship seems to have rose partly as a correlate of human brain evolution.

It is probably accurate to say that the tribal, nomadic social groups in which hominid and early human ancestors lived had a morality based on a combination of pragmatism and spirituality.

Their population was limited. As brains increased in size, it became possible to create more complex hunting tactics, toolmaking skills and to enhance migratory acumen. Placing high value on the contributions of each tribal member was a practical, survival requirement, that would have created a moral code reinforced by both secular reasoning and faith.

Yet no creature simply thinks its way to survival. Indeed, the separation of cognition and instinct seems artificial. There are numerous examples of population dynamics influencing behavior and thought without the organism being aware of the circumstances.

For example, as discussed earlier, in r-selected populations parenting styles become less urgent. It is as though the collective (quantum entangled?) mind of organism/nature "decides" that since high birth rates increase the odds for survival of offspring, parents need not be so attentive. On the other hand, in k-selected circumstances where the environment is less supportive, and with a lower birth rate the value of each child is increased. Each set of circumstances is based on the successful continuation of the species.

All creatures respond to these circumstances in similar fashion, but they do not 'think" their way to different parenting styles. Instead, there seems to be an influence on mind from nature that does not originate in consciousness. In that context the atheist (we don't need God) argument leaves a lot out; most notably that man cannot regulate himself as if a moral tabula rasa who must be taught how to be moral.

Also, a small tribal group will tend to have closer genetic/familial ties, which means their social morality will derive ultimately from their biology. Finally, the division of labor within the group would tend to preclude the kind of socioeconomic competitiveness and resentment seen with densely populated enclaves. Such a social model might appear to preclude behaviors we now regard as sinful. It is as if descriptions of Eden in the Bible were based on unconscious images of pre-urban, tribal society.

Despite pragmatic and nature-based moral elements, humans have typically carried out their moral obligations to please gods and spirits. It is as if the human mind was designed to believe in a higher power, and that once a volume/ function threshold in the brain was reached the need for transcendence and worship was effectively imprinted within us.

Notwithstanding the possibility of an ingrained need for transcendence and faith it also appears religion and society co-developed. When human beings from different genetic lines and cultures were forced to get along, morals, laws and religious concepts were required to ameliorate the duress and alienation accompanying those changes. With increased diversity, religion became more personal.

In that context, it is entirely possible that, despite the materialistic arguments of Marx and Engels, the ego-alien struggle to accommodate, live with and accept genetic strangers has been partly responsible for much of the political strife in human history. Without the religious accommodations inspired by figures like Buddha and Jesus things could have been much worse.

Have we met that struggle successfully, attained the socio-politico-religious golden mean, with altruism as a prevailing ethic even among strangers? Perhaps not. Wars based on cultural, ethnic, racial, and ideological differences have permeated human society since the first towns were set up along the Tigris and Euphrates.

Some of this might derive from our primate origins. When chimpanzees wander into the territory of another troop, savage battles break out. Obviously, that is not always true with humans, but only because we have two gifts enabling us to modulate, and channel aggression. One is language, which we use to negotiate, threaten, predict, and placate. The other is God. While in ancient times God "went into battle" - and, in a sense, still does in the Middle East, He provides us with reasons not to, via tenets like the golden rule, personifications like the Good

Samaritan, the idea of universal compassion, and the notion that all people are God's creations.

To presume we no longer need faith is rather like post-Freudian psychiatrists demeaning the contributions of Sigmund Freud when, to paraphrase from Carl Jung, they could only have attained their modern, post-Freudian insights by standing on his shoulders. Since there has never been a clear separation between God and man there has never been a time when Homo sapiens did not act in terms of a religious doctrine. Only later, when science emerged as an alternative means of control was religion deemed unnecessary in some quarters.

In the final analysis, there seem to be two compelling voices in the human mind commanding us to act in ways compatible with religion. One is the evolutionary draw of our social systems. By virtue of our primate ancestral roots, we will always gravitate toward social hierarchies. There will be idols, icons, superstars, political leaders, artistic geniuses, eminent religious leaders, and scientific ground breakers. For those in the "woke" generation who believe tearing down statues of iconic figures will level the playing field and produce a pure socialist, egalitarian society, the odds are against it. Political systems might be referred to as egalitarian, but Homo sapiens knows, in his heart, mind and genes that without some sort of hierarchy he is lost.

Human beings cannot escape the social rank order. The problem comes when we adopt human icons and leaders and find them inadequate or tyrannical. At that point where does Homo sapiens turn? It appears we cannot conceive of a perfect king or president, because there are no perfect people. All of us have an id complex. All of us are part of nature. Utopia is something we are strangely compelled to seek. That is the fundamental reason for politics, but there is no such thing.

Utopia is something we can only aspire to via worship and divine transcendence. Once again, there seems to be a central motive compelling worship of gods and/or icons locked into our neurology and shaping our fundamental beliefs.

The same paradox is involved in the notion of equality. We write about it, believe in it, but it is a concept that can never be realized. Nor do we necessarily want it. Consider the questions: What if there were no movie stars, super athletes, esteemed religious figures or political leaders to look up to? What if there were no heroes coming to the rescue, no prominent figures providing us with legendary accomplishments and heroic traits we can impart to our children? Would we survive psychologically?

Even the American founders knew this. Thomas Jefferson accepted the inevitability of the hierarchy. He espoused equality but his use of the word 'equal' pertained to basic rights and legal access rather than to equal talents, intelligence, drive, and motivation. In fact, his discussion on the natural aristocracy suggests he found nothing wrong with one person being more successful than another. It was the inheritance of wealth and status for simply belonging to a family that irked him.

The hierarchical mandate, combined with human imperfection, led the founders to place God firmly into the Constitution, the Bill of Rights and into the mores of the United States of America. God was needed, both to satisfy human need (and instinct) and to prevent any human from seeking absolute dominance.

Another reason God is needed in some form derives from the most recent version of the human brain. Enlarged frontoparietal circuits enable us, not just to observe and act but to imagine what could be - to plan and dream. To consider the future, means to anticipate and be able to control the future. If there are no concrete explanations as to what that future portends, the human mind, with its vast neural circuitry, will tend toward closure seeking. In so doing, the mind will come up with ideas, systems, and if necessary, transcendent beings.

Perhaps because of such multiple influences it seems throughout history there has been a close and cooperative relationship between mind and God. It is possible that in the truest sense of the word, they are interchangeable. This does not imply that man is a God or that God is an illusion, merely that the idea of God being found within us might be more than metaphorical. It might be entrenched in mind as an evolutionary adaptation to life's vicissitudes.

If so, certain questions come to mind. Is God above man, or an extension of man? Are both woven into nature's tapestry, making distinctions between them secondary to the integrative wonders of the cosmos? That question was never more relevant than during the twentieth century, when physics provided a plethora of new laws, relationships among particles and questions about whether man could separate himself from nature enough to observe it.

Amidst this came an intriguing notion called entanglement, which provided a new view of the universe - including the existence of a single, non-material entity hovering over all forces, actions, matter and human interactions.

CHAPTER 17:
ORDER AND UNCERTAINTY

THE NAMES COMPRISE A DREAM team of scientific discovery; Einstein, Planck, Heisenberg, Schrödinger, de Broglie, Bohr, Bohm, Feynman, Pen rose, Hawking. But it didn't begin with them or with advent of the twentieth century. Certainly, some of the greatest discoveries took place at that point, beginning with Planck and Einstein. However scientific discovery is a relay race and every advancement stems from a prior accomplishment.

Arguably, advancements of the 20th century were preceded by the work of Michael Faraday and James Clerk Maxwell. The latter was to the former as Newton was to Galileo. Faraday tinkered with gadgets and phenomena related to the movement of electrons. His insights and inventions were brilliant, but he was not able to describe them using precise mathematical formulas. It was left to Maxwell to do that through a set of calculations that came to be called (appropriately enough) Maxwell's equations.

One of the questions back in the late 19th century had to do with "what made the world tick." What made it move? What brought things together to create mass and energy? Mechanics knew there were forces that created power and speed. It was, after all, a time in which industrialization was spreading around the world. Machines were needed to produce materials, clear earth to build roads and carry cargo. The question was, what laws of physics made all that possible?

There was also great interest in what light was composed of - an understandable concern that could be traced back to the origins of the human species because of sunlight's life-giving properties. However, its essential nature and elements remained unknown. Isaac Newton concluded early on that light consisted of particles. His reasoning was based on the idea that light traveled in a straight line. He dismissed the alternative theory that light was a wave because waves feature undulations and directional changes. His description of light as a series of particles was rejected in many quarters, and for a time, wave theory prevailed. It would take a few centuries for that to be re-addressed but it would not be completely resolved, even up to the 21st century when ironically, the most accurate descriptions suggested it is both a particle and a wave. That paradoxical duality continues to shape scientific discourse in virtually every university physics program around the world. Still, there was a prevailing belief that there are features linking light to electromagnetism. Since questions on the nature of electromagnetism were more easily resolved it seemed a reasonable starting point.

In 1831 Michael Faraday discovered the process of electromagnetic induction. By creating interactions between magnetic fields and electrical currents he was able to conclude that they resulted from the same process. He also brought a new theory of light transmission to the fore. In what came to be known as the Faraday Effect, he found that magnetic fields could influence the pattern of reflected light. Suddenly, two important phenomena of concern to scientists, manufacturers, and lay persons - light and power - seemed to be created by magnetic interactions.

While Faraday's work had practical applications, its theoretical implications were even more profound. It started scientists wondering just how varied were forces within nature. Was it possible that some singular, galvanizing force was controlling everything; perhaps a mechanism embedded within nature, or perhaps an entity rising above concrete observation while being responsible for all universal phenomena? It stimulated interest in finding a grand unified theory that would grow stronger with time.

Faraday answered the question partly, but the process behind the electromagnetic phenomenon had yet to be determined. That's where James Clerk Maxwell came into the picture. In 1845, he developed equations describing the nature of electromagnetism on light, including the factors of charge, speed of transmission (through a vacuum) and the proportionate relationship between specific variables. It was now becoming clear that components of the natural world previously deemed separate were perhaps part of a larger "oneness."

While the 19th century came and went, the search for an ultimate integrative explanation of how nature (and the universe) works continued. Among the first concerns for 20th century physicists was the nature of energy. Through the work of Faraday and Maxwell it seemed two phenomena that drive the universe (light and energy) were the result of fields. Just what was meant by that, either then or even today, is difficult to tell. A field is presumed to represent a continuous spread of matter and energy that engulfs the objects over which it has influence. The closest thing to that is a wave, which has continuity of "flow" but breaks up eventually as its energy subsides.

Newton's idea of light as a particle had been on the back burner until 1900, when German physicist Max Planck came up with a novel idea that moved science light years ahead. He determined that all particles, molecules, and atoms obeyed a central principle. Rather than energy spreading through the sweep of fields, it consisted of discrete packets, and that nothing could either emit or absorb energy except through interactions among those packets. Even more important was his proposed "bottom line" measuring standard. Planck concluded that energy had limitations on value, that is, an absolute lower limit to the amount of energy that could be transmitted. He also discovered a proportionate relationship between the frequency of photon (light particle) transmission and the energy photons could convey. This came to be called Planck's constant and the packets he referred to came to be called quanta. His idea set the stage for an area of exploration known as quantum physics.

The implications were stunning, not just because of Planck's description of nature's most basic elements, but because his ideas led to confusion about the most microscopic elements of the cosmos.

Previously, it seemed like the field explanation was valid. If not, it would be difficult how separate electrons, neutrons, protons end up acting as though organized into a chain. If each had its own energy quantity and limitations as well as being disconnected from other particles, what was the "glue" that created continuity among them, and for that matter, in nature itself? In other words, how did the chain reaction linking particles create force and energy?

Though brilliant, it seemed as though Max Planck's discoveries harked back to a quaint theory proposed by Greek philosopher Zeno regarding the impossibility of movement. Zeno had abstracted his way to a belief that an arrow thrust forward did not move in the common use of the term. Instead, he proposed that each thrust consisted of a discrete step, followed by another step, and that these steps were physically separate, with no cohesive, fluid connection. Zeno had little credibility in his time and obviously has less so in modern times. Yet, Planck turned out to be right about quantum theory - sort of.

He had his followers. One was Albert Einstein. During his magical year of 1905 he solved a lot of problems in physics and went on to revolutionize how we think about space, time, light, energy, and existence itself. His first venture came with the discovery of the photo-electric effect. With this, he resolved some of the uncertainty resulting from Planck's discovery.

With respect to light and energy, Einstein borrowed from Newton by suggesting light consisted of particles, which he called photons. He offered a solution to the question of how light packets could travel like a wave, while still consisting of discrete quanta. In an experiment that resulted in his receiving the Nobel Prize, Einstein bombarded a sheet of metal with photons. He found that regardless of the volume of photons bombarding the metal sheet no electrons were released from the metal sheet unless the photons reached a high frequency wavelength. This was curious. Since photons convey energy, one had to ask why an accumulation of photons did not always energize and release the electrons from the captivity of the metal sheet?

Einstein's explanation was as follows. The red frequency of light is lowest, the blue frequency is highest. The energy factor was not a result of how many photons were sent but the speed at which they travel. Thus, any amount of the blue frequency photons would cause the release of electrons from the metal sheet, whereas no amount of red frequency photons would. The fact that speed of transmission was the causative factor in energy transmission demonstrated that blue frequency photons traveled fast enough to make individual (discrete) photon packets appear to be connected and wave-like. It was much like the effect on the eye of a motion picture that seems to move fluidly from one scene to another, when in fact it is separate film fragments running at a rapid pace from clip to clip.

The earliest theories of electromagnetism suggested continuous light waves transferred their energy to electrons. Einstein's experiment disconfirmed that notion. It also bridged the gap between wave theory and quantum theory. It turned out the wave-like appearance of light was due to a swarming of individual packets that in some instances would reach a high frequency threshold. It wasn't a question of waves creating energy but rather of individual high frequency photons doing the honors. In other words, a collection of photons could seem like a wave to an observer due to their speed and the perceptual limitations of the observer.

Discovery of the photoelectric effect still did not answer the question of how photons communicated with one another (i.e. line up in cooperative/cumulative fashion to create a wave effect). Even if one high frequency photon could cause release of an electron through energy infusion, that would not explain how a series of photons could increase the energy effect, particularly since they are discrete quanta. Was it a one at a time, cumulative effect, or a weaving together of discrete particles into a more powerful energy force? Einstein's later work on light as a constant would provide an answer, which was in one sense natural, in another, metaphysical.

Why the natural/metaphysical terminology? As Rudolfo Clausius discovered, all things in nature obey the principle of entropy, that is, the idea that every system eventually decays, or runs down to a state of disorder. The amount of energy remains the same, as per the law of conservation, but nothing remains in its original form through time, with one exception - the photon.

The photon does not decay. Some particles have incredibly long half - lives but all eventually decay. Light does not. The reason why is both simple and complex. It is complex because the process by which decay usually occurs is through the weak force, which releases radiation that "wears things down." Just why light is exempt from entropy has not been determined but there are some possibilities.

At light speed, time does not lapse for the photon. As Einstein demonstrated, time slows down as one approaches light speed. At the point where an object is traveling at 186,262 miles per second, time stands still. As a result, the photon cannot abide by temporality. For example, if a photon begins at point A, then continues to point B, one could assume it took time to reach its destination. However, while a human observer would view this in temporal terms, the photon itself will not have aged or experienced a time-lapse. That is because light speed (c) cannot be surpassed.

For any object to reach the point of entropy requires a time-lapse. The weak force is time bound, thus cannot exert its influence on the photon. Radiation is a cumulative process, and decay is a function of time. To use another example: an infant will not develop wrinkles, gray hair, and arthritis until a substantial amount of time passes in its life. However, if the infant was traveling through life at light speed, it would never get older. What does this say about light? Is it material or does it exist in some other domain? If the latter, did the Egyptians have a point in assigning God-like status to the sun? Questions about the nature of light speed and time would turn so confusing that scientists would eventually resort to mystical explanations about how this might pertain to the origin of the universe. Various explanations were proposed. Some involved the timeless existence of multiple universes. Others offered the idea of unification between man and nature through the anthropic principle. Einstein was instrumental in resolving such quandaries.

Photons fascinated Einstein. He eventually discovered that light was the only constant in the universe and that it was also a bit mysterious, because while it had a specific rate of speed it also had no rate of time. Rate implies a temporal transition - an acceleration. Yet, he also discovered that light does not speed up or slow down. Its speed remains the same, even, incredibly, if its path is warped by gravity. Although it can bend it cannot change its speed. It has a regulatory purpose in the universe - as if it were a quasi-physical manifestation of transcendence.

Einstein eventually turned to another topic that was equally vexing for scientists. Having lent a helping hand to Planck by taking quantum physics to the next level. He next turned to Newton to provide help in dealing with an old concept known as gravity.

Newton knew how to measure gravity (more or less) but he didn't know how it worked. Therefore, his attributing this phenomenon to God's creation made perfect sense. Einstein was able to insert nature into the process because he possessed an unusual type of mind that was part scientist and part dreamer.

The brain of Homo sapiens provides the capacities for mechanical reasoning; primarily through spatial, holistic circuits in the right cortical hemisphere. It also provides sequential, grammatical, symbolic capacities through the operations of the left hemisphere. Many people favor one skill set and hemispheric function over another. That doesn't mean mechanical types are deficient in the areas of language expression and logical thinking. It does mean people who discover they are proficient in one of those two areas will often engage in tasks that are easier for them, have more practice in those skill areas, thus leading to slanted development in one of the hemispheric skill sets.

Einstein seems to have been equally proficient in both areas. As a result of such cognitive versatility, he was able to not only solve complex math equations but also use his versatile intellect to engage in "gedankenexperiments"- research of the imagination. In so doing he figured out how gravity worked. On one hand, his theory presented an incredibly sophisticated exercise in cognition. On the other hand, it was akin to an imaginative exercise seen in child's play. Consider the following vignette.

Imagine three people in a room. There are two older gentlemen and a young child. The men are seated across from one another at a table, while the child is under the table playing with a little ball. The men are discussing gravity, comparing notes, hoping to come up with an explanation for its existence. As they compare opinions, alluding to historical ideas on the subject, the child continues to roll the ball back and forth on the floor. He hears the discussion above him heating up.

Frustration mounts as the men hold heads in hands asking: "Why do objects fall from the sky?"

During this, the boy notices that the ball tends to roll faster in areas where the floor is slanted. The wooden floor is a bit warped in places and the ball follows the curvature of the floor. At that point the lad yells out to the men above: "Things fall from the sky because the sky drops down in some places. The sky is tilted in some places - like the floor." The men initially laugh at this but then realize that, of course, the universe is uneven in places, that it does "drop" in a sense and has a fabric-like composition that makes smaller objects fall toward larger objects. With that realization they finally come up with an explanation of gravity.

Einstein conceived of the universe as a spatial fabric and demonstrated that gravity was the result of geodesic geometry. Place an infinitely vast sheet of cloth around and through the cosmos, put a variety of large and smaller objects on that cloth and sure enough, the small objects will tend to roll toward the larger objects. It was a superb melding of the right and left hemispheres in producing a seminal discovery.

Despite his groundbreaking discoveries, Einstein had himself a dilemma that was twofold. First, his theory of gravity was based on a steady state universe. It was a model conceived by Fred Hoyle, who described the cosmos as kinetic, but poly-stable, which meant it had fluctuations here and there but did not change its net shape or composition. The second reason was due to a new discovery that put man's position as the sole creature capable of uncovering the secrets of the universe in serious jeopardy.

Whether Albert Einstein truly believed God created the universe or was merely presenting a contrary argument to Werner Heisenberg's principle of uncertainty is hard to tell. He most definitely believed the cosmos was orderly, that with sufficient understanding of its laws one could determine how it began and how it would end, if that was its eventual fate.

However, something called the double slit experiment cast doubt on the idea of a lawful universe. The existence of laws means events can be predicted if sufficient variables are known. The point where an object starts out, its course and its endpoint should be knowable providing its mass, speed and directional origin can be determined from the outset.

However, that basic tenet of logical, empirical thinking was sideswiped by a rather mundane but mysterious experiment first conducted by Thomas Young in 1801; later replicated in 1927 by Davisson and Germer. Young used photons, which made the results perhaps more predictable due to the non-temporality of light. Since light is its own, in some ways marginal, paraphysical entity that did not necessarily challenge the idea of a lawful universe. However, the Davisson-Germer study used electrons, which do obey the constraints of time. That really put a fly in the ointment of an orderly theory of the universe - alternately referred to as the realist school and classical physics.

It began with photons being run through a gathering apparatus with a single slit. As would be predicted from the particle (quantum) theory of light, the path of the photons could be measured with accuracy. However, when a second slit was added and there were two possible paths to the gathering board, something strange occurred. The photons did not pass through the slits predictably. Instead of a specific endpoint involving only two possibilities - where one photon went here and one went there, a smearing effect, also called an interference pattern, occurred. The photons' path and momentum became scattered and only predictable using probable outcomes. In other words, there was only a chance it would end up in one location or another.

Interestingly, while its initial starting point could be determined its endpoint could not - and vice versa. That seemed to suggest it had no regular path and simply did not obey the laws of motion, and of physics per se.

At first this seemed illusory, but the experiment was repeated many times with the same results. The final conclusion was that at the smallest levels of existence, i.e. the particle world, the cause-effect model so critically important to scientists in understanding nature, simply did not apply. This became even more confusing when electrons were used because while the photons' non-temporal behavior could be attributed to the fact that they did not have to be in any specific location or reach a particular endpoint (a travel sequence requiring a time-lapse), the photon could understandably cheat in the experiment. However, since the use of electrons yielded the same result, it createdenormous confusion.

Since the result was duplicated many times, it was finally decided that the photon and light itself acted like both a wave and a particle: a particle when it had only one possible path, and a wave when it had leeway to "choose" between more than one path.

All this aroused suspicion and frustration. The field of physics was divided. Some, including Einstein, felt there must be a hidden variable to explain why this occurred; one that agreed with the lawful (classical/realist) view of nature. The other school maintained that the uncertainty principle was valid even if there was no explanation behind it. The discovery seeped into the quantum physics domain.

New theories emerged. One held that the reality that we think of is somewhat illusory, that the classical model of cause-effect doesn't apply at the subatomic level. In the aftermath of experiments dealing with this phenomenon, Danish scientist Neils Bohr bought into the uncertainty principle, stating at one point that we must accept the fact that the universe is not lawful.

Einstein bristled at this and developed a contentious relationship with Bohr. At one point the former expressed his disdain for uncertainty in the famous quote: "God does not play dice with the universe." But there was more to come. In subsequent experiments, it became clear that both the location and momentum of particles could be determined when the researcher was observing the photon's path. In other words, it appeared the particle was playing coy with the researcher: taking paths according to mere probabilities when the "teacher turns her back", but, like a clever class clown, "behaving itself" by taking predictable paths when being observed.

The main explanation for this was that when a person observes something, he is emitting photons through his visual sense and is therefore interacting with the experiment rather than merely observing the particles. In other words, the scientist, being part of nature and necessarily emitting photons from his eye while observing, was a causal factor- an obligatory participant rather than neutral researcher. The implication was that there is no such thing as detached observation, that we cannot study nature, only interact, and communicate with it. When we change, it changes and vice versa....an example of apparent "cosmic oneness."

This put scientists in an awkward position. Confusion around this problem was intense and the phenomenon came to be called "quantum weirdness."

Despite Bohr's concession to the uncertainty principle, less compliant scientists continued to address the problem. It seems the dual nature of subatomic matter was too incongruous to simply accept. Why did photons act in two different ways? Research on the photoelectric effect demonstrated that they are particles, yet the uncertainty principle demonstrated that they acted like waves - because only waves produce interference patterns.What gives?

Several explanations came out of this, including Louis de Broglie's pilot wave theory. He proposed that particles and waves are both present in not just photons and electrons, but all objects in nature. His compromise model suggested that as an object's mass increases, the wave function diminishes, until with large objects the wave property is canceled out.

The reason for the decline in wave influence with increased mass is that as the mass of an object increases so does its gravitational pull. At a certain level of mass, gravity reins in the wave function until eventually the wave is completely consumed. It seems to suggest everything in the universe, including humans, lions, tigers, and automobiles, functions like a mini black hole.

de Broglie's theory was an attempt to explain why the classical laws favored by Einstein and the laws governing the quantum particle world could both be valid. The idea had flaws and was never proved, but neither were most alternative theories dealing with differences between the classical and quantum versions of reality.

One of those alternatives was fascinating because it clearly extended beyond the domain of empirical science. It was based on the notion that a photon travels along every possible direction, along every possible path, before landing at a specific endpoint. This was Richard Feynman's 'Sum of Histories' theory. At face value, it might sound strange, except for two supporting features. First, the photon has no temporality. Because it does not obey the restraints of time it does not obey the restraints of space. (Note that Einstein's work proposed a dimension known as space-time, whereby the two were effectively co-dependent). By Dr. Feynman's calculations, the only way to determine the path of a photon or particle (electrons seemed to behave similarly in the double slit experiment) was by calculating the probabilities of its position and momentum as it passed through its myriad paths.

The idea that timeless, non-spatial features were built into the cosmic fabric led to speculation of an even more fascinating theory called entanglement. In this model, physicist David Bohm suggested there was no true separation among objects in the universe, and that the universe was very much like one giant electron, as though a gigantic tree with branches representing differentials in mass, distance, force, and charge. This theory suggested the apparent distinctions between stars, planets and all celestial boundaries are illusory. One impetus for this ultra-singularity theory (my words, not Bohm's) was due to growing interest in another theory known as non-locality.

In looking out at the universe over enormous distances, it is possible to determine the temperature and mass of various galaxies. It seems, while there are plenty of bumps and grinds along the way, the universe is rather uniform in terms of those features. If there was an origin to the universe - whether caused by a big bang or some other event, one would expect there to be massive differences among various sites.

As an analogy: consider a single area in the USA – the Nevada-California border. Parts of this area are hot desert, while other parts are cold, even during the same time of year. The temperature in crossing the border from Reno, Nevada to Squaw Valley, California will shift from 90 degrees, with hot, arid conditions, to freezing temperatures and blizzard conditions along the mountain roads.

In that context, the sheer amount of cosmic matter spread far and wide should feature significant differences in the temperature and mass of the cosmic terrain. However, that is not true. It appears the universe has a smooth topography. Even more perplexing is that for this to be true means the communicative influence among distant bodies in terms of temperature and mass would not be possible without some built-in nexus. For instance, while pouring warm water into a glass containing ice cubes would eventually cause a flattening out (compromise) in the temperature of the water, that would not be the case if there were two separate glasses of water, one with ice cubes, the other with just warm water. The two water temperatures would have to interact to reach a flattening effect.

Because of cosmic uniformity, the argument was made that the universe really is one large, interwoven fabric-like structure. That's where the idea of entanglement comes into play. Entanglement theory assumes the universe is one "something," and the fact that mere observation can affect the results of an experiment means all creatures, including man, are part of that entanglement. Does that mean the capacity to determine causes, effects and differences by the human mind are false capabilities? Theoretical physics has been moving in that direction for decades.

Many explanations have been offered to explain the uniformity of the cosmos. Some have been devoted to finding a theory of everything. Some examples are superstring theory, brane theory, and holographic theory. The problem is that theoreticians seem to have run out of explanations, at least within the rubric of traditional methods of scientific inquiry.

Some theories emerged, not from observation or probabilistic prediction but from abstract mathematical concepts that are not derived from anything physical. Some require the existence of up to eleven dimensions. Some, such as M theory and its derivatives, suggest there is a parallel universe separate from our own by only a few millimeters, and that the proximity of one to the other caused a big bang-like collision now presumed to have started the present-day universe.

That model holds that, unlike the original big bang theory, nothing had to be created from scratch. The material essence of "multiverses" is presumed to be constant. It entails a fascinating parallel to the original conception of Fred Hoyle's solid state "net-universe" but while arguably "retro" it does not solve the problem of time and material origin. Interestingly, despite being so replete with energy and so perilously close together the multi-verses are only presumed to collide every 15 billion years.

Such ideas seem unprovable, especially if one holds to the idea - now ingrained in physics - that this is an anthropic, all-in-one world where the observations of man are woven into the tapestry of the cosmos.

If the anthropic principle is valid, it must be considered that human powers of observation have been defined and limited by our need to survive in a specific planetary environment. We cannot see or hear in eleven dimensions. Nor will we ever be able to determine if another universe is running alongside ours. Physicists can draw conclusions by observing energy signatures and creating particle collisions in enormous underground tunnel complexes, but that always attaches to known facts about the variables being studied. What if, in the quest for a unified theory, we discover that we don't have sufficient brain capacity to achieve this goal? Where will we turn? How will we satisfy our curiosity when all the interesting theories of the cosmos must remain just that - theories.

The field of theoretical physics has provided substantial progress since the start of the 20th century. Einstein hoped to find a unified theory but was so, frustrated that he once declined to present a paper on the subject commenting that the idea he was about to describe would have been the greatest blunder of his life.

What will humans, inclined to seek closure, do if no further resolutions can be found? It isn't as if we haven't traveled down this road before. Homo sapiens has always had questions about nature. Our survival has always depended on our understanding of nature. When concrete answers were lacking, we found other solutions. Pantheism, paganism, monotheism, philosophy, metaphysics, and theosophy were all cognitive survival tools that kept us going. They weren't artificial, foolish, or mere defense mechanisms against the anxiety fomented by an uncertain, dangerous world. They worked. And we survived and still exist because of them.

Over time, ideas and methodologies have changed. However, one constant concern has revolved around the question of how it all got started - how *we* all got started. From the earliest human societies there were stories of creation. That has not changed with time.

CHAPTER 18:
IN THE BEGINNING

THERE HAS ALWAYS BEEN GREAT INTEREST in the search for origins, mostly because we are drawn to the notion of beginnings. Tantalizing questions revolve around what hominid line led directly to us, what combination of forces and events led to the formation of the earth, what entity existed before the big bang, or perhaps in an artistic context, what antecedent musical forms branched off into modern jazz. One could ask why humans have such a fervent need to know how things got started. One possible answer lies in the nature of the human mind.

The search for origins is part of the closure process. To understand a beginning is to understand why subsequent events occurred. Such knowledge also serves the purpose of reducing uncertainty within the vast human brain, which explains why humans in virtually every culture conjured up a theory of creation.

Among the first recorded stories of creation was the Babylonian epic, Enuma Elis. According to that text, the world was forged during a battle among the gods. Since the story has the Babylonian God winning out, this myth probably derived partly from political concerns. Just as the gods were fighting, so were the real-life armies of Babylon and Assyria. It was an instance in which myth imitated life.

The story of creation in the Old Testament bears similarities to this but also features significant differences. Both accounts include an initial period of chaos, specifically an encounter with a flood. The main difference is that the Hebrew version revolves around establishment of a monotheistic faith. While creation is preceded by chaos in that text, it is not related to human combativeness. Obviously in a monotheistic context, there could be no battle among competing gods to determine how the world was formed. Rather, it was an internal battle between pride and humble submission to God. Despite that, what stands out in many accounts of creation is a process whereby disorder precedes an orderly endpoint.

The number of creation myths is staggering. Whether it be the Ainu myth of Japan, the Hopi legend of Tawa, the Egyptian story of Geb and Nut, most versions begin with some sort of cataclysm. In one account, there is no land for man to inhabit, so birds are sent by the creator to splash and patch their way to solid ground. In another account, the first act of God is to seed a lifeless sea into fertile land through transformations. Other myths involve the creation of man, followed by God's decision to flood the earth without saving all its people.

A central theme seems to run through most of these stories, which might reflect not only the functions of the human brain but also a critical feature of the natural world. As discussed above, the human brain evolved as, among other things, a noise breaker. Its volume enables us to store enormous amounts of information, but that same volume can create high levels of noise, such that sifting through the neural morass to find responses and memories can be difficult. Indeed, without the cataloging functions of grammatical language we would be rather overwhelmed by the input volume we are able to process.

This is exemplified by non-verbal autistic individuals who experience chronic anxiety, hyper-emotionalism, and cognitive confusion, and require extreme repetition and structure to function on a day-to-day basis. In evolution, humans adapted to a noise heavy brain by developing both language and a compelling curiosity drive, designed to seek out and pre-resolve conflicting inputs to prevent being overwhelmed by sudden confusion.

Those cognitive mechanisms are also responsible for most of the wonderful outcomes of the imagination.

Another factor derives from information theory, which proposes that everything in nature operates through a communicative process, and that all signals involve some initial level of uncertainty. Indeed, information is said to be attained only with a reduction of uncertainty – with each reduction comprising one bit of information. Indeed, as Shannon and Weaver demonstrated, information content can be measured with precision.

All accounts of creation seem to reflect the same chaos-to-order directionality that governs human brain function. In that context, it is not surprising that scientific accounts of creation also begin with chaos and end with order.

Scientific theories of universal origin changed with time and new discoveries. Einstein believed the universe remained in a fixed state, that what we observe is the final product. Despite his ambivalence about the existence of God, he suggested a single governing entity made and governed the world. Fred Hoyle proposed that the cosmos did undergo changes but remained in a net "steady state."

However, those ideas were challenged by Edwin Hubble. He understood that variations along the light spectrum reflect differently, depending on whether an object is approaching or moving away from the observer. When approaching, its frequency slows down. The red spectrum is the slowest frequency and objects will reflect red as they become closer. Conversely, the fastest frequency is in the blue spectrum, which is reflected when an object is moving away. When Hubble looked out at the cosmos through a highly advanced telescope, he observed blue spectrum phenomena. Celestial bodies seemed to be flying off - as if in the aftermath of an explosion. The big bang theory resulted from this and subsequent observations.

Not everyone came to accept this theory, largely because a massive explosion would not result in the relative smoothness of matter and temperature that is seen across the universe.

Physicist Alan Guth provided an explanation of how matter and temperature could end up dispersed uniformly to create smoothness. His explanation was called inflation theory. The idea was that there was an expansion so rapid that there wasn't enough time for normal, widely scattered dispersions to occur, it could explain the contradiction between the big bang theory and the isomorphic quality of the universe. This theory has not been proved but has gained wide acceptance.

The problem with all theories is the difficulty determining how much an idea describes nature objectively and how much emanates from the functional dispositions and biases of the human brain. While there is a big difference between telling a story of creation through oral traditions and drawing conclusions from cumulative research, there is still the possibility of a hidden variable influencing both science and myth. One hidden factor could be the co-functions of nature and the neuro-psychological biases of the human brain. To their credit, modern physicists don't deny this, as seen in their acceptance of the anthropic principle.

As with mythical accounts, there have been differing explanations of origins within scientific circles. One theory suggests there was no single big bang event or actual beginning to the universe. Instead, advocates of M theory have proposed that adjacent membraned universes sit aside one another and that at some point they cross an infinitely small cosmic border and collide, thus creating a tumultuous event before the universe settles down. It is a strange concept, arguably unprovable. it does solve the problem of material and temperature uniformity, of origins, and of the beginning of time. However, such phenomena might never be observed.

In reviewing the various ideas of creation, it seems even some of the brilliant minds involved in cosmology might be flailing a bit. While there is certainly merit in arguments presented by renowned thinkers like Dr. Stephen Hawking regarding the absence of a true beginning and the pointlessness of religion, man seems trapped in a corner, pulled on one hand by his oneness with the universe, and on the other by the tentative belief that he can truly be an objective observer and empirical scientist.

It appears the very individuals who seek objective, non-theistic explanations of how things work have been forced to submit ideas founded on the existence of eleven dimensions, an infinite number of universes that somehow emerged like bubbles out of a cosmic boiling pot, or a holographic universe that doesn't manifest itself until observed by the one species capable of integrating its fragments.

Humans cannot at this time observe such hypothetical phenomena, let alone prove their existence. By itself, that might not be significant. Many ideas in physics are derived, not from observation but from mathematical extrapolations. For example, superstring theory is a creative, integrative concept but one many physicists believe lacks the construct precision to be testable. Furthermore, many ideas now fully accepted have never been verified by observation. No one has ever seen an atom or an electron, let alone a quark. Their existence is inferred from energy signatures traced after interactions. An atom might look like a sphere, a square, or a banjo string (one reason superstring theory invites such fascination). Still, even if man is tied to nature so functionally as to be part of a larger gestalt, scientific inquiry will, and must continue.

Is there some resolution to all this? Is there some concept that can be said to underscore all phenomena in nature; a process responsible for the motion of an electron, a father's love for his daughter, the speed of light and the behavioral and moral requirements needed to maintain faith and social order?

Moreover, can one create a conceptual bridge between nature and God, particularly one not personified, but an amorphous force overseeing all operations of the natural world and nature?

There is one possibility. It is a theory of everything, embedded in all objects and actions, from gravity to altruism, from the process by which one electron influences another to a mother's reaction to an infant's first words. It is a curious principle that is both abstract and mathematically precise. It can be measured but not seen. It is not material or strictly spiritual, but it permeates all aspects of life. It has transcendent qualities that can meet the criteria of many religious systems, because it is formless but ubiquitous, creates order from chaos, provides answers to perplexing questions about life, humanity, and the prime functions of the cosmos. It was once described by Werner Heisenberg as the theory that decides, and it is discussed next.

CHAPTER 19:
SIGNALS

JOHN ARCHIBALD WHEELER had a vision. It was in response to irresolution arising from various, increasingly esoteric theories of the cosmos. Quantum mechanics did not solve the problem of field unification and it seemed every attempt to explain the operations of the universe in terms of its non-local, probabilistic structure seemed to come up short. Bottom line: Wheeler and some in the realism school of thought believed Einstein's classical concepts work better in the world we live in, even if quantum physics remains useful.

Many of today's technological gadgets are developed in accord with quantum principles. On the other hand, if you want to send a satellite to the moon and are not necessarily interested in the spacecraft taking all possible historical routes before getting there, classical physics seems to work better.

Still, physicists have refused to divest from either classical or quantum models of the cosmos, so a group of iconoclasts added something called quantum information theory, which integrates quantum physics and information theory. Out of this came the idea that the universe might operate like a computer i.e. as an information system based on the binary principle. That sounds a bit more complicated than it really is. 'Binary' simply means every action, every decision, every event ultimately involves one of two possible answers - a yes or a no. In binary terms a 1 or a 0.

The idea was regarded in some quarters as revolutionary, in others as a tautology. Of course, events play out like that! Either the big bang occurred, or it didn't. Either a photon will energize an electron, or it won't. Either Mary will say yes to Joseph's marriage proposal, or she won't.

Theoreticians were skeptical about the possible application of this model to cause-effect dynamics. In other words, they questioned how one might use this idea to determine why events occur. It lacked the bite of a theoretical model containing an independent variable.

John Wheeler tried to fill these gaps by assuming every system in nature has a fundamental feature that cannot be broken down further. This was his version of Democritus' atom, of Green and Schwartz' superstring and of Peter Higgs' boson (also known as the God particle, because it was assumed to act as a field in which force was converted to mass).

Wheeler called his new idea "it from bit" and it signified that all things in nature arise from resolutions between two or more indeterminate components. He captured his thinking in the following way:

"It from bit" theory espouses that in essence every item of the physical world derives from an immaterial source and explanation, that what we call reality arises, in the last analysis, from the resolution of yes-no questions and registration of those resolutions. The idea is that all things physical are information-theoretic in origin and this is a participatory universe."

At the risk of putting words into Dr. Wheeler's mouth, this passage provides an integrative bridge between pure science and the existence of a non-material entity in control of all things, as if tantamount to a God - not a God personified, who oversees the plight of a specific group of people but a non-material process that does not actively control everything but does determine everything. It is not part of something else, like one member of a trinity, one part of an exchange of quarks, or one celestial body having a reciprocal gravitational relationship with another. It stands alone. Interestingly, while deriving from advanced scientific concepts, it seems to represent a turn back on the road to religion, particularly pantheism.

Without offering a challenge to Dr. Wheeler, whose intellect and knowledge surely dwarf those of this writer, I would like to apply elements of information theory to a broader range of topics, including all human experience, and apply it to an understanding of the nature of God. First, some basics of Information theory.

The theory has components that have been applied to physics, astronomy, communications, mathematics, biology, genetics, and psychology. In order to discuss its effect on human cognition, scientific inquiry and nature requires some discussion of those components. A clear understanding can be gained by describing four of its main features. These are...

Communication/Interaction
Uncertainty/noise reduction
Encoding
Systemic integrity.

With respect to communication; as John Wheeler wrote, everything in nature occurs and/or is formed through some sort of interaction. There is no solo act in the cosmos. The interaction can consist of electrical charges between two particles, between water rushing over a rock to create erosion, or to a wife's request to her husband to take out the garbage. It is a restatement of Newton's third law of motion, which states for every action there is an equal and opposite reaction. In this instance, it means no effect can occur unless one stimulus provokes another. This means we live in a world that is nothing but a communications network. That is the clay from which all persons, objects, beliefs, and forces are created and sustained.

However, there is a snag in that process. For an interaction to create an effect, information must be conveyed. What is information? In terms of information theory, it is defined as a reduction in uncertainty (uncertainty can be used interchangeably with 'noise,' chaos and entropy). There must be some degree of Disorder for information to be obtained. In other words, information has no a priori existence. It must be extrapolated from some prior amount of disorder.

That means we live, not only in a world of interactions, but also in a world in which the baseline default position is disorder. Perhaps that is why the human brain tends toward what Karl Lashley called mass action and proceeds from general activation to specific memory retrieval or response selection.

Lashley applied this to various aspects of mental activity. For example, when we attempt to figure things out - whether solving a problem in algebra or picking out a pair of running shoes, more neural systems are activated than are needed to access a specific sensation or memory. In other words, general arousal comes first. After that, the brain receives feedback relative to the parameters of the problem and a sifting process streamlines mental activity. Amidst all this, responses, memories, and perceptions are formed.

As one might expect, the brain operates exactly like the natural word in which it evolved; from the general to the specific and from neuro-psychological chaos to order by reducing uncertainty. It is an information machine.

Information can be measured, though that is not relevant to this discussion. For example, before flipping a coin the outcome is uncertain. The fact that one does not know which side will come up means there is uncertainty. It could be heads or tails. However, once it is flipped one possibility is eliminated. Whether the toss ends up heads or tails, one bit of uncertainty is reduced, which means one bit of information is obtained.

The words entropy, disorder, and chaos (which, as stated earlier, can be used interchangeably) refer to uniformity and an absence of distinguishing codes signifying that some parts are separate from others in terms of structure and function. All systems in nature undergo entropy. We often think of this in terms of the aging process. However, in addition to experiential aspects of getting older, are the information aspects. In death, the organs that once functioned separately, albeit for the central purpose of fueling the cells, lose their individuality. The heart, lungs and muscular system cease their distinctive functions and become monotonized, non-functional tissues. In death, the body completely loses its information content as well as its homeostatic vibrancy. To be alive means having biological information content - an unusual way to look at life and death, but it is nonetheless valid.

In addition to communication, uncertainty-entropy and coding, another aspect of information theory has to do with systemic integrity.

Encoding distinct structures and functions to reduce uncertainty and increase information does not guarantee continuation of the system. As discussed above, all systems tend toward entropy, including those previously encoded. There are two ways nature can overcome that. One is by adding a periodic influx of energy - for example sunlight reinvigorating plants in spring. Another way is by increasing information content by making an entity more complex to provide it with a check and balance, as well as a redundant homeostatic mechanism that insulates it from decay for longer periods of time.

The latter occurs when subsystems crop up that make distinct contributions to the overall system but are nonetheless partially separate from the system. (An example is found in the complementary relationship between bacteria and the gastrointestinal tract of most organisms). With increased complexity, an entity can change without coming undone. Having vast information content, whereby subsystems are encoded/distinct but functionally and structurally connected to the main organism, enables the overall system to maintain stability.

It takes more uncertainty/disorder for a complex system to decay. The human body presents a classic example. The prime purpose of life is to provide nutrition to the cells for energy storage and expenditure. If a body had only one organ by which to do that a small injury or infection could destroy the entire system. However, with more than one organ system and therefore greater redundancy, harm to one organ system would not necessarily destroy the overall entity because the others could kick in to sustain homeostatic balance.

The human body has multiple ways of providing fuel to the cells. Gastrointestinal, cardiovascular, pulmonary, and musculoskeletal systems operate in integrative fashion, which improves physiological resilience, and by the same token, preserves the information content of the body. In other words, life span is closely related to organic complexity. Once the process of complexity is underway the tendency is for the entity to endure. It is a process typically referred to as poly-stability.

Information theory can apply to the origin of the universe. For example, if, as some maintain, the temperature of the pre-expansion universe was too hot for atoms to form, there would have been no distinctive (coded) elements (molecules or atoms) within the "cosmic egg." That would have had enormous ramifications. Because mass is needed for force, an entropic (uncoded) universe would have had no mass. That means it would have no force. In addition, because force is needed for movement there would have been no movement. Finally, since according to relativity theory movement and space are correlates, there would have beenno space.

That means the cosmic egg was both an entropic "something" yet nothing because it had no information content. It had to obtain information to originate systemically because the law of energy conservation prohibits any manifestation of energy out of the blue. Yet it was not systemic. It had no capacity for communication since all elements were fused and non-interactive. In essence, the rule of thumb in information theory is that if an entity is not encoded it does not really exist.

Eventually, the cosmos became informed and separate elements were able to interact. Perhaps this happened when the temperature declined a bit in some places, allowing matter to form. That first code/ distinction would have led to the origins of energy, force, space, and time. But it was not likely the big bang or any other single event led to the origin of the universe. It was the accumulation of information content, when separate structures and functions cropped up that began to interact and cooperate systemically. That led to cosmic expansion, and stability through the influence of gravity.

As with all things in nature, the origin of the universe involved a rapid journey from uncertainty (or entropy) to information. While this would seem to contradict the notion that all systems in nature run down, it does appear logical, especially if one assumes there was no system prior to an epoch of cosmic Information. In that context, it would appear the most significant provider of information (as well as being a force) was gravity.

Due to the interactive nature of gravity, the cosmos became functionally unified so that effects in one area tended to influence those in another area. In some instances, this involved embellishment, in others cancellation. In the aftermath, there were massive explosions everywhere. Celestial bodies came and went, and while these volatile entities were more informed/encoded than in the original cosmic egg, the universe was still less informed and systemically poly-stable than it would later become. Once gravity exerted its influence, galaxies and solar systems became more distinct, complex, and systematized. To determine how this could apply to religious worship and the nature of God, it might help to engage in a brief review of information theory.

Entropy, disorder, and chaos are states without distinctions or codes. It is a state of uncertainty (or lack of information) that applies to all things in nature. While every system will eventually run down, entropy can be postponed when there is an added influx of energy to replenish its distinctions and functions. We see this in every aspect of nature; for example, when (as alluded to earlier) dormant plant life re-emerges in spring through solar reinvigoration, when novelty is added to a stale marriage or mundane job to re-create interest, when genes are reshuffled through sex so there is some degree of variation from one generation to another, and when a musician decides his pop music style can be enhanced by changes in sound, arrangement and lyrics.

Each interaction, whether occurring in an interpersonal exchange or a cosmic reaction, contains some noise that interferes with the message. In order to obtain information, there must exist a prior state of uncertainty. In other words, information cannot exist on its own, instead must be extracted from noise. That means the laws of classical physics necessarily derive from those of quantum physics rather than the two being mutually exclusive. While physicists typically argue over the relative validity of classical and quantum models – as if only one can be valid, information theory would hold that quantum mechanics provides a seedbed of uncertainty from which the codification of the cosmos (uncertainty reduction occurs – in other words they are complementary. In that context, it appears Information Theory is indeed the theory that decides.

While the relevance of information theory to the fields of physics, communications and cosmology seems clear, its influence in other areas of life requires a more in-depth explanation. The question is whether it can be considered a prime social moral factor and meet the criteria as a morally governing entity. In this writer's opinion it does meet those criteria.

John Wheeler suggested, information dynamics regulate all natural phenomena. In the domain of physics, there must first be a distinction between two particles so that they can affect one another. In a moral context, a similar dynamic prevails. The Ten Commandments describe and prescribe acceptable practices in terms of a monotheistic religion. However, in the shift toward worship of one God, it was necessary to distinguish one faith from another. Judaism had to be encoded, as did Jews themselves, through dietary, theological, and social distinctions.

God was presumed to appeal to a distinct group of people. Just as the universe could only have arisen when material distinctions occurred. that is, become encoded with information content, so too was it necessary for the Jewish religion and its adherents to distinguish themselves in various ways for their monotheistic faith to prevail. That information dynamic can be applied to all religions, including Christianity, Islam, and pagan and pantheist systems. Some systems are informed by distinguishing between and among gods based on name, purpose, and origin. While having so many deities could have led to a diluted faith in these religious systems the stories and myths are woven together to create poly-stability.

There is, in all religions what St. Augustine referred to as synthesis and divergence. For example, Greek gods are said to come from single family lines - as with Zeus, Apollo, Hera etc. That is an encoding process that gives both novelty and complexity to a single religious concept. In other instances, a central belief holds the variations together. This is not a perfectly fluid process, however.

In the Islamic and Judeo-Christian faiths - as well as with other religions, there are conflicting elements. In some instances, there is a message of compassion, of respecting and even protecting foreigners and strangers. In other instances, strangers are subject to cruelty, neglect, and abuse.

How does this square with information dynamics and the systemic integrity of each religion? In Islam, Judaism and Christianity poly-stability is attained through obedience to the will of God. There is, in all three faiths, a prime commandment that holds true across all events, trials and tribulations. That allows and compensates for the inevitable vicissitudes that will occur over time. It holds Catholics together, even after eating meat on Friday and divorce are sanctioned by the Church. It held Jews together, even when they experienced doubt about the existence of God during the holocaust. And it keeps the Islamic faith intact despite the antipathy between Sunni and Shiite. God is, in effect, a code holding humanity (a species heavily influenced by ancient impulse) together. However, while it seems clear information dynamics provide systemic maintenance, their relevance to basic moral principles, including determining right andwrong is a complex issue

As discussed earlier, it is difficult to separate moral principles from evolution-driven human behavior patterns. For example, each of the commandments serves to enhance human survival by prohibiting behavior patterns that disrupt social cohesion and by favoring those that enhance solidarity. In both biological and religious terms, transgressions like stealing, coveting, killing, and bearing false witness would create discord, alienation, and internecine violence.

The cross-influence between evolution and religion is obvious. The evolutionary sequence from archaic to modern human occurred mostly when humans were living in relatively small groups. The responsibilities of hunting, gathering, weaving, exploring caves, signaling the presence of rivals or predators were all crucial in sustaining not just the group but the genetic line.

Human groups most capable of maintaining solidarity among members who exhibited those various skills would have more likely survived long enough to pass on their genes.

How does that pertain to information dynamics? One way is through encoding within the population, i.e through individuation and establishment of identities. Without such distinctions the human group would deteriorate – something seen in communist systems where individuality is driven out by a collective mentality.

More concretely, consider, if all members within a social group had the same degree of skill, strength, the same height, weight and level of attractiveness, social entropy would occur. That has evolutionary and religious significance. Since it took a variety of skills to enhance survival among early humans having diversity within the population would have been adaptive. That means, like the universe, both human anatomy, and human society had to become increasingly informed by having both synthesis and divergence to prevail and thrive. Novelty/encoding was needed for invigoration and synthesis was needed to keep the system integrated.

However, it wasn't just about individual skills and traits. Groups of any sort require leadership. Since the genomes of early humans incorporated primate genetic patterns, there would have been a need and tendency to establish a social hierarchy, perhaps not in the sense that existed later in monarchic, urban settlements, but in terms of social rank, which would likely have been carried out in conciliatory ways due to small population, tribal altruistic dynamics. In line with the information content of the group, some would be leaders, others, followers. Some individuals might be forceful communicators, which would have helped them keep the peace when conflict arose and motivate others in times of duress.

The social encoding process would have utilized information dynamics to improve the group's chances of survival. Just as encoding solidified the group so would it have solidified belief systems that provided a common, unifying frame of reference and source of motivation, persistence, and hope.

Morality can be discussed in information terms because it is a function of norms. Moral behavior is what everyone thinks they are expected to do. As to the question of how moral tenets formed in the first place, morality probably has its roots in biology.

While Sigmund Freud insisted the component of the human psyche devoted to primal instincts (the id) had to be either suppressed or channeled to accommodate the need for social and cultural cohesion, many primal behaviors are still considered moral; for example, maternal protectiveness, altruism, social cooperation and even jealousy, which, despite its negative connotations solidifies the commitment of parents to a stable family dynamic and facilitates child rearing.

However, instinct also derives from information dynamics. Behaviors in any given society (both prosocial and antisocial) are encoded within that society. They become symbolized, accepted as norms. At any point in time deviations from the norms might be considered immoral. But since morality operates as an information system, it will require some degree of novelty from time to time to prevent entropy. In other words, even with respect to morality, stagnation will tend to become entropic.

That means what was once deviant can become acceptable; not due to political trends or even out of necessity, but because the human mind, like everything in nature, operates in terms of information dynamics. Society needs a replenishment of energy lest its values decay. Whether one believe the transition from the staid 50s to the hyper-liberal 60s was a beneficial trend, some sort of cultural transition was probable necessary.

Such trends have occurred throughout history. Sex before marriage was once taboo, but in the 60s it became more acceptable, until by the eighties, remaining a virgin past early adulthood was virtually a mark of shame. Slavery evolved along similar lines. It existed in every part of the world. American, Chinese, Japanese, African and European societies all engaged in the practice. Consequently, it was viewed by many as a normal and necessary feature of an existing economic system. People of all races, ages and genders were involved in the slave trade and there was expressed justification for the activity. For example, some maintained since there was no public education in most places, slavery at least provided food and shelter for people with no substantial means of supporting themselves It was not a black or white thing - as of 1776 there were still over fifty white slaves in Boston, Massachusetts. Also, European serfs during the Middle Ages were almost exclusively white.

However, thankfully, morals shifted, partly because more populous, merit-based governmental ideas were being proposed by Thomas Hobbes, Jean Jacques Rousseau, and John Locke. There was also a shift in thinking about various races and cultures. Native Americans did not know much about white Europeans or vice versa. They interacted, some even married across racial lines but by and large estrangement prevailed.

The same was true regarding African slaves. Being born in an African culture not yet industrialized, they were burdened with a presumed image of inferiority. The first wave of slaves did not speak English and there were no formal educational opportunities by which they could learn the language. In many instances they were not allowed to be educated for fear that increased knowledge might lead to a boost in confidence and self-worth, political enlightenment, and subsequent revolt.

Some, like Thomas Jefferson were less fearful of revolt and more optimistic about their potential. Jefferson was surprised at the intellectual abilities of native Americans he came across and those of his slaves. His relationship with Sally Hemming, a slave, was both affectionate and consultative, as demonstrated by his bringing her to Paris during his tenure as ambassador to France. In a letter to William Short he even expressed the opinion that had slaves and "Indians" been exposed to "letters" (his term for reading material) they might have produced a Voltaire or a Racine.

Thus, the accepted norm of slavery, reliant as it was on the belief that enslaved (black or white, male or female) were in that position due to inherent inferiority was changing as a result of new egalitarian philosophies, particularly through the writing of John Locke, whose tabula rasa theory proposed that all people are born "blank slates," and that experience and learning, not in-born ability determined one's life course. As soon as the new code (equality) was in place, many began to view slavery as an abomination. And so, the information encoding process continued.

The it from bit concept has fascinating implications, both because of its breadth and its applicability. It can be discussed in terms of the origin of life. For example, the usual argument regarding biological origins is between those who disagree on whether RNA and DNA or protein molecules came first. The RNA/DNA argument holds that the most essential factor in the origin of life was a mechanism for reproduction. Others insist amino acid-chains of protein came first, due to the need for tissue formation and catalytic reactions that produce energy and ultimately a metabolic system.

It is possible neither of those points is accurate. As with the universe, the origin of life might have resulted from reaching a threshold of information content. Consider that, even if DNA and protein components existed in the primordial soup (the earth's early tumultuous atmosphere) it might not have mattered. Billions of years ago the earth's climate was volatile - very hot in daytime, very cold at night. While bio-friendly chemical compounds could assemble, they would usually break up due to those climatic extremes.

Biochemicals couldn't have assembled into an intact system if not for a stabilizing entity separating and insulating them from the outside environment. In other words, the protein and DNA/RNA macromolecules had to be encoded, that is, separated structurally and functionally from the outside environment before there could be systemic stability.

Bio-ecological uniformity (entropy) had to be broken up for life to originate and most importantly, endure long enough to create a long-term succession of generations. Thus, the argument can be made that life became separated from the prebiotic world by the emergence of a barrier/insulator in the form of lipids, which were protective membranes encapsulating both tissues and reproductive materials. Just as the universe had to be encoded to exist, so did life. Such is the breadth of the information-based argument

John Wheeler was criticized when he came out with it from bit theory, not because it was wrong, but because it was in some ways too correct. Additional criticism was based on the view that, while information theory was descriptive, it had no real deterministic value and did not address the issue of causation or rebut criticism that while everything in nature adheres to its principles it is not in itself an interactive variable. It is not quantum/discrete or classical. Nor does it consist of a field of energy. It describes how things happen but does not prompt occurrences. The question is whether those are valid criticisms. One way to approach this question is through use of this author's use of a gedankenexperiment.

Imagine that everything in nature is the same in every respect, that each star and planet has precisely the same mass, that there is only one kind of particle and that all humans are exactly alike with respect to their physiology, genetic makeup, and behavior patterns. Also, imagine there is only one season, and that the temperature is the same every day. Finally, suppose there is only one speed in the universe - light speed.

What would occur in those circumstances? First the entropy arising from mass equivalence would eliminate gravity. The sun could not attract the earth if both had the same mass. That doesn't mean celestial bodies would fly off, however because there would be no disproportionate gravitational pull in any direction. Also, with no differentials in mass, there would be no space warp because that can only occur with mass differentials. In effect, there would be no planets or stars because, due to ultra-uniformity, they could not communicate or interact with one another. That would prevent the influence and existence of force. In effect there would be no interaction, no existence, and no systematization of the cosmos.

Now, further imagine all particles are the same, regarding charge, mass, and color (referring to the spin of the particle on its axis). The particles would not really (functionally or structurally) exist because they would be completely inert, almost like shadow versions of active particles.

Next, imagine there was nothing in the cosmic egg but sameness, that even the hot plasma featured no differences in viscosity or temperature. In that case, no part of the primordial entity could influence any other because there must be a distinction between one component and another to create a communication - a message, even if it's nothing more than a minute energy differential.

Now take it to the social level, including the domain of religious worship. There would be no such thing as personhood if all attitudes and behaviors were identical. Nor would there be any possibility of communicating. Ultra-redundancy would lead to social entropy and preclude the sending and reception of signals. Even with a voluminous brain there could be no interpersonal perception, no likes or dislikes, no sense of better or worse, no big, small, beautiful, or ugly. Nor could there be any distinction between morality and sin.

What would occur in such circumstances? At face value, these examples might seem absurd, because it would suggest it is possible for there to be existence without existence, form without substance, people without people. On the other hand, a research project by Von Bekesy demonstrates how such a 'to be/or not to be.' scenario could be valid.

In conducting an experiment on visual perception. Von Bekesy flashed visual images at the retina of his subjects, and they responded in a way indicating they saw the stimulus. He then taped open the eyelids of his subjects to prevent eye blink responses. That served to create an almost uniform visual field and precluded input renewal.

Eye blink responses continually change the visual field. They do not interfere with the retina or lens, but they do create shifts in the field to override sensory monotony (and uncertainty). In the experiment, subjects became functionally blind when eye blinks were prevented. Despite having their eyes open, a fully functional retina, optic nerve, and intact visual cortex they could not see.

The implication of this study - now well substantiated, was that movement in the lids and change in the visual orientation was as necessary as the functions of the visual organs. In effect, perception required a reversion to uncertainty followed by a reduction of uncertainty. Without information dynamics there could be no such thing as vision.

To all intents and purposes, the images flashed before the subjects with taped eyelids were both there and not there. Their existence depended on information content available to the viewer. That in turn required the extraction of information from uncertainty through the periodic re-codification provided by eye blinks.

While eye blink research and cosmology are different animals, the same principle applies. The existence of everything in nature depends primarily on its information content.

Since this book is concerned with the religious/moral implications of that, one could ask whether religious figures and belief systems can exist without information content being part and parcel of the process. It is difficult to answer, because like information theory itself, the issue is rather abstract.

God has had so many faces and names throughout human history that it is difficult to adopt one model as a frame of reference. He or She is not an independent variable. God had no name or face in the Jewish faith, indeed, to pretend to see him or use His name was considered blasphemous. Presumably that was because Hebrews needed Him to be all powerful and ubiquitous. If He could do anything, be anywhere and everywhere, it would have been absurd for Him to have physical attributes. The act of placing Him in a specific time and place, with physical dimensions would mean He could not be everywhere.

Most religions assigned some sort of unreal, non-physical stature to God, even the Greeks, whose gods had traits above that of mere humans. Thus, for the most part, the human concept of God has precluded shape and substance. That is significant.

If there is a tendency for the human species to look to amorphous entities for guidance, hope and worship, then perhaps it is not unreasonable to assume information theory could be considered a modern version of a supernatural entity because it governs everything, and nothing can exist outside its parameters. It cannot be experienced directly but it can be acknowledged because it is part of all aspects of life. It is more than a theory of everything. That is concerned only with matter, force, and chemistry. Information dynamics extend well beyond that, influencing physical laws, social relationships, art, language, cognition, brain function, homeostasis and even politics.

As discussed earlier, it also has bearing on faith, morality and the nature and existence of a divine entity that governs everything, with status above that of humans. The parallels are clear. God is said to be within us - we are created in his image. That is also true of the information bit. Its dynamics are built into our brains, our language, our physiological systems, our politics, our math, our marriages, our genes, our cultures, our mechanical creations, our planet, our solar system, and our galaxy. If nature is entangled, there must be a connection between mind and nature. If our species is inclined to seek hierarchies to function and if one aspect of that search for protection, nurturance and guidance comes in the form of God, then perhaps one can assume that experience is part of the picture as well.

British physicist Roger Penrose had an interesting take on entanglement. As discussed earlier, this refers to the fact that experiments have shown that particles separated by time and space can interact (or at least correlate behaviorally) with one another. Entities separated by distances that seem unable to interact because of the limitations of light speed can still exhibit mirrored, simultaneous changes parallel to one another.

In fact, one difference between an information - based concept of the universe and the oneness implied in entanglement is that information requires both oneness, (pan-connectivity) and distinctions through the encoding process. It is those two factors thatprovide resilience for all things.

In a sense, entanglement theory and the quest for a unified theory would seem to contradict information principles. If one assumes entanglement is the natural state of the cosmos, effectively connecting everything within a single fabric, with no inherent separateness or information content, entropy would be the result. In that context, one could argue that the logical endpoint of entanglement theory is entropy, i.e. nothingness rather than oneness.

Meanwhile, Information theory allows for the continuation of systems through distinctions within systems via the encoding process. In effect, the rule of existence on every level in the natural and spiritual world seems to be twofold. There must be coherence (a degree of redundancy or "oneness" and a degree of change or distinction.

We know from common experience that everything is not connected. It if were, the cause-effect model that successfully describes nature would be useless. It provides a conundrum that some have acknowledged. For example, Roger Penrose attempted to integrate the quantum (particle) and classical/realism ideas by proposing that there is a component of the universe without barriers or individualized entities. In that state the universe is indeed one entity. However, there are interactions among objects causing that oneness to break up - that the universe is, as Wheeler suggested, an information system in which there are continual shifts from stability to encoding (detachment/differentiation) among its elements. Penrose called this phenomenon "observation" and presumably referred to the double slit experiment in quantum physics whereby observations by researchers seem to change the outcome of the experiment. His words might be rephrased to mean "interaction"...which is a more physically pristine, inanimate description of the process. It essentially means the law of the cosmos is based on communication, that the universe is, above all else a language medium.

This has religious implications because God is seen as a totality. His influence spreads across the cosmos, influencing everything. Yet His influence can be broken up by interactions. In the case of human faith, those interactions include the variances of honor and sin.

It might be highly speculative, but one could surmise that through a neuro-experiential vista, human beings "know" there is continuity in nature, that all things are intertwined to an extent, and that there is a pure state of existence (akin to paradise) where nothing is broken or disrupted, where the entirety of existence is egalitarian, and disruptions that deviate too much from systemic beliefs can contaminate that continuity. Perhaps we recognize this in both a conscious and natural way. We deem it sinful or immoral when such entropy-inducing deviations occur, for example excessive pride, parochialism, and selfishness. Is it possible all our attempts to describe the nature of God, and to require an encompassing "uber-regulator" are at a fundamental level, the same, regardless of religion or philosophy? Is there an innate moral template analogous to Noam Chomsky's theory of innate language that does not derive from culture? Is it possible that there is a systemic "mono-law" governing everything from the laws of physics to the laws of society which consists of a prime commandment holding that there must be unification periodically infused with re-energizing distinctions enabling humans to endure and that this process is interaction-based?

One could rely on information dynamics to make that argument, especially in terms of the social-moral tenets of various religions. When we sin by disrupting and threatening the integrity of the social group, when we treat our own bodies in ways that jeopardize maintenance of its homeostatic balance, we act in defiance of its laws and veer in the direction, not of hell's condemnation, but of entropy.

Does that make information an amorphous quasi-deity? If so, should mankind build temples and create rituals on behalf of the bit?

Perhaps that would be not only unnecessary but rather silly. For one thing, the gods we worship now are already part of that information system - no need to convert to the bit. For another, if the notion of an Info-God was valid it would have to be a God espousing both unity and the individuation/encoding process that comes with liberty and free expression. It - rather than He or She would be a dualistic, deistic idea/entity designed to preserve all systems, physical, biological, and spiritual through an encoding and resilience process. Because of the sheer range of that God concept, rather than expressing blind devotion, we would simply need to recognize its breadth to be filled with wonder.

In earlier times, the Greeks tried to reason their way to God. Aristotle sought answers to ultimate truths and the synthesis between nature and God. If he were around to ponder the insights of Claude Shannon, Werner Heisenberg, and John Wheeler he might have had his own insight, perhaps stating to his perplexed, but enthralled students at the Lyceum...

"Eureka! The thoughts I had, the words I wrote, the stylus with which I composed Politics, Physics and Metaphysics: all that - all of Stagira, all of Greece, all the heavens – everything part of a whole, yet with distinction among its parts. The sun, the moon, the ether, the animals, and all mankind, unique but woven together, not by a punitive, distant God but by an informative and imperceptibly communicative God, who seeks to hold it all together. At last: The Unmoved Mover! "

REFERENCES

Afterlife in Judaism: Article in Jewish Virtual Library Dec. 2020

Aiken, C.F. Redford, D. The Monotheism of Akhenaten, Nov 8, 2021 BibleOdyssey

Aquinas, T. (1981) Summa Theologica. New York; English DominicanFathers.

Armstrong. K. (1993) A History of God: A 4,000 Year Quest of Judaism. Christianity and Islam. Thrift books

Aronson. E. The Social Animal MacMillan Education Books) 1972)

Ashley, K.S. (1929) Brain Mechanisms and Intelligence, University of Chicago Press

Assman, H. (January 1997) Moses the Egyptian; The Memory of Egypt in Western Monotheism. Cambridge, Mass. Harvard University Press.

Bachvarova, M.R. (2016) Sargon the Great: From History to Myth. Chapter 8, in: From Hittite to Homer; The Anatolian Background of Ancient Greek Epic. Cambridge University Press 166-198

Berlyne, D.E. (1960) Conflict, Arousal and Curiosity, New York, McGraw Hill

Bernstein, I.S. 1976, Dominance. Aggression and Reproduction in Primate Societies. Journal of Theoretical Biology. Vol. 60, 2,7 459-472

Bhaumik, M.L. (2017) Is Schrödinger's Cat Alive? Quanta 6 (1)

Bishtawi, A. (Aug, 14, 2017) Origin of Arabic Numerals; A Natural History of Numbers. Author House

Blair, P. (Dec. 2013) Reason and Faith: The Thought of Thomas Aquinas. The Dartmouth Apologia.

Bohm, D. (1980) Wholeness and the Implicate Order. Routledge

Botulis, A. Jagger, G. (2010) Philosophy of Quantum Information and Entanglement. Cambridge University Press

Bowersock, G. (1996) The Vanishing Paradigm of the Fall of Rome. Bulletin of the American Academy of Arts and Sciences Vol. 49 (8) 29-43

Boyer, C.B. (1985) A History of Mathematics. Princeton University Press

Broughton, J. (2002) Descartes' Method of Doubt. Princeton University Press

Brown. L. (2005) Feynman's Thesis. World Scientific.

Carlson, D. G (2007) a Commentary to Hegel's Science of Logic. Pal grave MacMillan

Catani, M. Dell'aqua, F. Vergani, F. Malik, F. Hodge, H. Prasun, R. Valabreque, R. Thiebault, M. (2011) Short frontal lobe connections of the human brain. National Library of Medicine: Cortex. 48 (2) 273- 291

Cellos - Vasquez, V. P, Garcia-Domingez, F. Martinez, M.A. (2009) Unusual High Frequency of Hermaphroditism in the Monophonic Bivalve Journal of Shellfish Research 28 (4) 785- 789

Clinton, B. (1998) In Search of Muhammad. Continuum International Publishing Group pp. 182-183

The Complete List of Royal Families of Ancient Egypt. Thames andHudson, (2004)

Dann, M. (July 13, 2018) The Essenes and the Origin of Christianity, article in the Jerusalem Post

Davies, W.W. (1905) The Codes of Hammurabi and Moses, Cincinnati, OH Jennings and Graham

Dawkins, R. (1976) The Selfish Gene Oxford University Press

deBary, W. (1969) The Buddhist Tradition in India, China and Japan. Vintage Books

Dimitrov, M. Nakic, M. Waxman, J.E., Granetz, J. O'Grady, J. Phipps, M. Milne, E. Logan, G. Hasher, L. Grafman, J. (2003)

Inhibitory attentional control in patients with frontal lobe damage. Brain & Cognition 52 (2) 258-270

Doak, R.S. (2005) Galileo; Astronomer and Physicist. Capstone

Dorman, P.F. (August 2012) Akhenaten; King of Egypt. Britannia Encyclopedia

Doorman, B. (Nov. 2021) Plato and Aristotle; How Do They Differ? Britannica

Dunbar, R.M. (The Social Brain Hypothesis and its Implications for Social Evolution 2009 Annals of Human Biology)

Dunbar, R.I.M. (2010) How many friends does one person need: Dunbar's number and other evolutionary quirks. London, Faber and Faber

Eaton, O.N. Simmons, V.L. (1939) Hermaphroditism in Milk Goats Journalof Heredity Vol. 30 (6) 261-266.

Ehrman, B (2020) Heaven and Hell: A History of the Afterlife. Simon & Schuster

Ellis, A. 2004) Rational-Emotive Behavior Therapy; It Works for Me. It Can Work for you. Prometheus Books.

Evans III F.B. (1996) Harry Stack Sullivan - Interpersonal Theory andPsychotherapy. London, New York, Routledge

Faye, J. (1991) Neils Bohr; His Heritage and Legacy. Correct Kluwer.Academic Publishing.

Festinger, L. (1962) A Theory of Cognitive Dissonance. StanfordUniversity Press

Fortin, J.R. (May 2008) Saint Anselm and the Kingdom of Heaven: AModel of Right Order. The Saint Anselm Journal

Fleck, S. J. (1995) Aristotle in the Medieval Era Internet Article

Francis Lord Bacon, (1605) The Advancement of Learning; The OnlineLibrary of Liberty

Freud, S. (2002 - reprint) Civilization and its Discontents. Penguin Press

Freud, S. 1928. Humor. International Journal of Psychoanalysis (9) 1-6

Fuster, J.M. (2002) Frontal Lobe and Cognitive Development, Journal ofNeurocytology, 31, 373-385

Gaukroger. S. (1995) Descartes; An Intellectual Biography. Oxford.Clarendon Press

Gumbel, S. Nov 9, 2020. Mitochondrial Eve: The Mother of All HumanBeings. Great Courses Daily

Golding, J. Genesis. (2020, Baker Academic Press)

Gowlett, J.A. (2016) The discovery of fire making by humans: a long and convoluted process. Philosophical Transformations; The Royal Society, Biological Science, 37

Guth, A.H. (1997) The Inflationary Universe; The Quest for a New Theory of Cosmic Origin. Basic Books.

Guthrie, W. K.C. (2020) A History of Greek Philosophy; The Earlier Pre-Socratics and the Pythagoreans. Abe Books

Harare, Y. Sapiens: A Brief History of Humankind (Harper Collins 2015)

Hart, G. (2000) A Dictionary of Egyptian Gods and Goddesses, London, New York, Routledge

Hawking, S. Penrose, R. The Nature of Space and Time. Princeton University Press

Heather, P. (2006) The Fall of the Roman Empire: A New History. Pan Books

Heisenberg, W. (1983) Encounters with Einstein and other essays on people, places and particles. Princeton University Press. p. 183

Hill, R.A. Dunbar, R.I.M. (2003) Social network size in humans. Human Nature 14 (1) 53-72

Hitchens, C (2009) God is Not Great: How Religion Spoils Everything. Swift Reads

Human Evolution: A Timeline of Early Hominids: Tree of Life, Life ScienceJune 2021

Jastrow, R. 1981, The Enchanted Loom. Touchstone Books

Kaiser, D. (2005) Physics and Feynman's Diagrams; American Scientist 93 (2) 156

Kellerman, K. Reynolds, R. (1990) When Ignorance is Bliss: The Role of Motivation to Reduce Uncertainty in Uncertainty Reduction Theory. Human Communications Research, 17 (1) 5-75

King, B. (2007) Evolving God; A Provocative View on the Origins of Religion. Doubleday Publications

King, B. (2016) Were Neanderthals Religious? 13.7 Cosmos & Culture; Commentary on Science and Society

Klaushofer, A. (2016) Pagan Advent; A Time of Secrets and Silence; Nature/Animals/Religion/Spirituality

Koenigs, M. Motrin, J. (2017) Psychopath's Brains Show Differences in Structure and Function. University of Wisconsin Madison Research Study.

Kraft, H. Dec. 1, 2000) Max Planck; The Reluctant Revolutionary. PhysicsWorld.com.

Lazarus, R.S. (1991) Progress on a cognitive-motivational relational theory of emotion. American Psychologist 46 (8) 819-834

Lewin, R. (1999) The Australopithecus; Human Evolution, An illustrated Introduction. Blackwell Science, 112-113

List of Angels in Theology, Wikipedia

Luria, A.R. (1959) The directive function of speech in development and dissolution, Word, 15(2) 341-352

Mark, J.J. (2016) Egyptian Gods; The Complete List - article in World History Encyclopedia

McEliece, R. (2002) The Theory of Information and Coding. Cambridge University Press

Mehra, J. (2001) (ed) Louis de Broglie and the phase wave associated with matter: The Golden Age of Theoretical Physics. World Scientific546-570

Miller, P.D. (2009) The Ten Commandments. Presbyterian Publishing Co.4-12

Mindel, N. (2021) Abraham's Early Life; Jewish History. Chabad.ORG

Misner, C.W. Thorne, K.S. Wojciech, H. (April 2009) John Wheeler, Relativity and Quantum Information. Physics Today 62 (4) 40-46

Moore, M. J. (2016) Buddhism and Political Theory Oxford University Press

Mulvihill, A. Annamares, C. Dux, P. Matthews, N. (2019) Self-directed speech and self-regulation in childhood neurodevelopmental disorders: current findings and future direction. Cambridge University Press

Nadler, S. (2015) The Philosopher, the Priest and the Painter; A Portrait of Descartes. Princeton University Press.

Note; Einstein had banked on a universal model featuring what he called a cosmological constant, i.e. a static universe. He realized that even aside from Hubble's discovery of an expanding universe this was incorrect, indeed eventually concluded that his own theory of relativity made a cosmological constant unnecessary.

Note; On Freud's medical model. Sigmund Freud's Structural Model ofthe Human Psyche. SCI HI Blog April 2020

Note: In Matthew 13:32-2. The Parable of the mustard seed draws a syllogistic comparison between the spread of mustard seeds- the smallest of all which, grow into the tallest and most nourishing, and the poor who, though seemingly insignificant in society can spread the word and by virtue of their allegiance to God grow in magnitude and inherit the earth.

Note; In Matthew 6:28 Jesus draws on metaphorical logic in comparing lilies of the field..."who do their work of drawing from the earth and are sustained by God"... to people (who are more important than lilies) yet also work, therefore would also be sustained by God.

Note; in response to the contention of quantum physicists Neils Bohr and Werner Heisenberg that at the most fundamental (subatomic) level nature operated according to probabilities rather than discrete measurable laws Einstein protested by stating God (The Old One) does not play dice with the universe.

Note; Phineas Gage was a rail worker who suffered damage to the frontal lobes in 1848 when a spike exploded and pierced the entirety of his frontal brain. He astounded the medical community when after several days he was able to resume working and appeared to function normally. While his social skills were affected in marginal ways his speech, motor and perception skills were intact. This event led to subsequent dismissal of frontal tissue as critical to elemental brain function and might have set the stage for frontal lobotomies in the treatment of psychotic and anxiety disorders.

Note; on Plato and the State. March 2009, Plato's Theory of Ideal State. Literary Articles

Nazi, R. June 2021, Account for the Strained Relationship between Saul and Samuel. Cegas Academy Lectures.

Oaks, J.A. (2014) Khwarizmi. In, Kalin, I. (ed) The Oxford Encyclopedia of Philosophy, Science and Technology. Oxford University Press.

Offer, K. (2017) Augustine on Heaven and Rewards; Knowing and Doing; A Teaching Quarterly for Discipleship of Heart and Mind

Pelizorri, E. Padovani, F.H. Perosa, G.B. (2019) Mother-Child Interactions: implications of chronic maternal anxiety and depression.Psicologia: Reflexao e Critica

Pennington, J. (January 2020) Jesus, the Philosopher. Biblical Mind

Penrose, R. (2012) Cycles of Time. Vintage Pub.

Pinera, A.R., Reznick, D. Bryant, M.J. Bashem, F. (2002) r and k selection revisited: The role of population in regulation of life historyevolution. Ecology 83 (6) 1509-1520

Pinker, S. (1997) How the Mind Works, W.W. Norton

Polanski, J. (1998) String Theory. Vol. 1 Cambridge University Press

Pribram, K. (1971) Languages of the Brain: Experimental Paradoxes andPrinciples in Neuropsychology, Brandon Books

Rickles, D. (2014) A Brief History of String Theory; From Dual Models to M- Theory. Springer

Rogers, C. (1980) A Way of Being. Boston. Houghton Mifflin

Rooker, M. (2010) The Ten Commandments: Ethics for the Twenty FirstCentury. Nashville. Tenn. B&H

Rose, H. Jennings (1991) A Handbook of Greek Mythology. LondonRoutledge.

Schoechel, D. (2012) The Story of King David in the Bible, JewishHistory, Biographies in Brief.

Sheppard, A.R.R. (1981) Pagan Cults of Angels in Roman Asia Minor,Atalanta 12-13, 77-101

Shopflin, K. (2007) God's Interpreter: The Concept of Celestial Beings.Ed. F. Vieterer, Tobias Nicklas, 198

Talbots, T. (April 23, 2013) Heaven and Hell in Christian Thought. Article in Stanford University of Philosophy Encyclopedia

Van de Meeroop, M. (2005) Hammurabi, King of Babylon BlackwellPublishers.

Von Bekesy, G, (1974) Some Biophysical Experiments from Fifty YearsAgo. Annual Review of Physiology, 36 1-16

Trotsky, L. (1962) Thought and Language. Cambridge, MIT Press

Watt, W.M. (1974) Muhammad; Prophet and Statesman. Oxford Univ. Press Article Retrieved 12/29 11

Welch, A.T. Mousali, G.D. (2009) Muhammad. In John L. Expositor (ed) The Oxford Encyclopedia of The Islamic World. Oxford University Press

West, M.L. (1971) The Cosmology of Hippocrates; De Hebdomadibus. The Classical Quarterly 21 (2) 365-388

Westfall, R.S. (1994) The Life of Isaac Newton. Cambridge University Press

Whitehouse, D. (2009) Renaissance Genius; Galileo Galilei and HisLegacy to Modern Science. Sterling Publishing.

Zeligs, D.F. (1954) Abraham and Monotheism. Vol 11, (3) 293-316